INTRODUCTION TO VALENCE THEORY

INTRODUCTION TO VALENCE THEORY

by

JEAN WORRALL, B.SC., PH.D.

and

I. J. WORRALL, B.SC., PH.D.

Senior Lecturer in Chemistry, University of Lancaster

AMERICAN ELSEVIER PUBLISHING COMPANY, INC.
NEW YORK   1969

© J. & I. J. Worrall 1969

American Edition Published by
American Elsevier Publishing Company Inc.
52 Vanderbilt Avenue, New York, New York 10017

Standard Book Number 444-19707-9
Library of Congress Catalog Card Number 74-99464

Printed in Great Britain

CONTENTS

Chapter		Page
	Preface	vii
1	*Evolution of the Atomic Theory*	1
	Atoms as Solid Particles	1
	The Electronic Structure of Atoms	2
	The Modern Theory of Atomic Structure	4
	The Dual Nature of Matter	4
	The Uncertainty Principle	5
	The Hydrogen Atom	5
2	*Atomic Orbitals*	7
	Specification of Orbitals by Quantum Numbers	7
	Relative Energies of Atomic Orbitals	9
	Accommodation of Electrons in Atomic Orbitals	9
	The Periodic Table	12
3	*Basic Principles of Valence Theory*	15
	The Ionic Bond	16
	Ionisation Potential	17
	Electron Affinity	18
	Lattice Energy	20
	Crystal Structure	20
	The Covalent Bond	23
	Shapes of Molecules	31
	Electronegativity	37
4	*The Valence Bond Theory*	39
	Graphical Representation of the Directional Properties of Atomic Orbitals	39
	Electron Density	45
	Orbital Overlap	45
	Hybridisation	49
	σ-Bonding	51

Chapter		Page
5	π-Bonding	53
	Resonance	57
	Delocalised π-Electron Systems	62
6	*Molecular Orbitals*	63
	Bonding in Diatomic Molecules	63
	Bond Order and Bond Strength	72
	The Three-Centre Bond	73
7	*Bonding in Noble Gas Compounds*	76
8	*Bonding in Transition Metal Compounds*	80
	Valence Bond Theory	81
	Electrostatic Crystal Field Theory	85
	Colours of Transition Metal Compounds	90
9	*Intermolecular Forces*	94
10	*Bonding in Metals*	96
11	*Hydrogen Bonding*	98
	Index	101

PREFACE

This book has been written to provide the student with an understanding of how results, obtained from quantum mechanical concepts, may be used to describe the bonding systems in molecules. Although the subject cannot be appreciated completely without the use of mathematics, many of the conclusions reached may be used to advantage by the non-mathematician.

An increasing number of introductory courses in chemistry include descriptions of bonding in terms of the simple valence bond and molecular orbital theories. There is scope therefore for a text which will satisfy the needs of students who do not intend to study chemistry to an advanced level on the one hand, and which will serve as a useful introduction to intending chemists on the other.

Chapter 1

EVOLUTION OF THE ATOMIC THEORY

We live in what is often called the 'atomic age', and most of us know something about atoms and the atomic theory. If you were asked to describe an atom of hydrogen, you would probably feel pretty confident that you could do so; but it is doubtful whether your description would satisfy the modern concept of an atom! Our ideas about the structure of atoms have undergone radical changes during the past century or so; and just as a scientist of fifty years ago would scoff at Dalton's picture of an atom, so the scientist of today finds that the ideas of Rutherford, or even Bohr, are now out of date. This does not mean that these great men contributed nothing to our present day knowledge—quite the reverse. Our ideas of atomic structure have been progressing over the years, and in order to understand the modern concept of an atom it is helpful to follow briefly the developments which have led to the latest version of atomic theory.

Atoms as Solid Particles. As long ago as 500 B.C., certain Greek scholars maintained that all substances are composed of infinitesimally small particles which could not be further divided; and they called these particles 'atoms'. This was all very well, but a theory is not of much value unless it can explain hitherto inexplicable phenomena, or be put to some other use. It was not until 1806 that the 'atomic theory' was of any practical value, and then John Dalton proposed some ideas which revolutionised chemistry. He suggested that atoms have the following properties:

1. Elements are made up of atoms which are indivisible and can be neither created nor destroyed.
2. The atoms of a given element are identical and have the same weight and the same chemical properties.

3. Atoms of different elements combine with one another in simple whole numbers to form molecules.

The third of these properties was the most important for chemistry, and it gave a basis to the development of the idea of valency. Subsequent findings have led to modifications of the first and second of Dalton's proposals: the existence of isotopes for example, and the discovery that atoms can in fact be split into even more basic entities.

Dalton's concept of atoms as minute solid spheres was torpedoed by J. J. Thomson, who discovered the electron in 1897, and by Rutherford, Geiger and Marsden, whose work led to the discovery of positively charged *nuclei*. These discoveries led to the first successful theory of atomic structure.

The Electronic Structure of Atoms. First, let us establish the three so-called particles of which atoms are composed. They are the *electron*, the *proton* and the *neutron*. The electron is negatively charged and has a mass of $9 \cdot 11 \times 10^{-28}$g. The proton has the same magnitude of electrical charge as the electron, but it is positively charged; its mass is $1 \cdot 67 \times 10^{-24}$ g (approximately 1836 times as great as the electron). The neutron possesses no electrical charge and has a mass nearly equal to that of the proton.

The first electronic model of the atom was proposed by Ernest Rutherford in 1911. He suggested that atoms consist of a positively charged *nucleus* surrounded by a system of electrons; the whole being held together by electrostatic attraction. Most of the atom is empty space, and the greater part of its mass is centred in the nucleus—which is small compared to the effective volume of the atom. In fact, Rutherford pictured atoms as miniature solar systems held together by electrical rather than by gravitational forces.

In 1913, Niels Bohr put forward the theory that electrons move in circular orbits about the nucleus. An atom of hydrogen, therefore, could be pictured as shown on page 3.

The most important postulate that Bohr made was that *only certain electron orbits are allowed*. For an electron to remain in a stable orbit, the outward force exerted by the moving electron—which tends to throw it further away from the nucleus—must

[Ch. 1] *Evolution of the Atomic Theory* 3

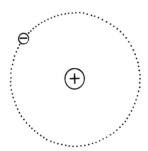

be exactly opposed by the electrostatic attraction between the electron and the nucleus. There is also a small gravitational force of attraction, but this is negligible compared with the electric force.

Let us visualise a positively charged nucleus surrounded by allowed electron orbits 1, 2 and 3:

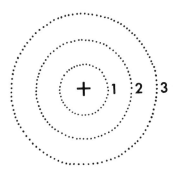

Then it can be seen that an electron in the orbit nearest to the nucleus (orbit 1) will be held more firmly by electrostatic attraction than it would be if it were in orbit 2. Similarly, an electron in orbit 2 will be more secure than it would be in orbit 3. If an electron is to move from orbit 1 to orbit 2, it will require some energy to pull it farther away from the nucleus; conversely, if it is to move from an orbit of higher number to one nearer the nucleus, energy will be given out. The electron orbit nearest to the nucleus will therefore have the *lowest energy level*, and the one electron in atomic hydrogen occupies this level in its most stable state. The most stable electronic state of an atom is called the *ground state*.

The Bohr theory was widely accepted, and can be used to explain many physical phenomena. It was later modified by A. Sommerfeld (1868–1951), who suggested that electrons could move in elliptical as well as in circular orbits. A further refinement was introduced in 1925 by Goudsmit and Ulenbeck who postulated that, in addition to being able to move around the nucleus, electrons may be considered to spin around their own axes either clockwise or anticlockwise:

The neutron, discovered in 1932, was also incorporated into the Bohr–Sommerfeld atom; and the presence of neutrons in atomic nuclei accounts for the occurrence of isotopes.

The Bohr-Sommerfeld theory was the first quantitatively successful atomic model, and it can still be used to explain a large part of chemical theory. There seems little point in dwelling upon it here, however, since it has been superseded in recent years by a far more powerful theory based on the wave properties of matter.

The Modern Theory of Atomic Structure

The Dual Nature of Matter. Until the beginning of the twentieth century, it was held that there were two quite distinct physical entities: *matter* and *waves* (or energy). Matter is made up of discrete particles (electrons, protons and neutrons); it is not continuous, in other words it is quantised. Waves on the other hand were traditionally considered to possess continuously varying frequencies and wavelengths.

This neat division of physical phenomena into groups of matter on the one hand and energy on the other was pushed away by Max Planck, Albert Einstein and Louis de Broglie. They advanced the irrefutable hypothesis that matter and energy are interconvertible; energy, as well as matter, is quantised.

When we come to consider a particle as minute as an electron we can no longer regard it as being distinct from a wave; and it is not surprising, therefore, that the laws which describe large systems in our everyday experience do not apply to particles such as the electron.

The Uncertainty Principle. The fact that electrons can exhibit the properties of both particles and waves gives rise to what is known as the uncertainty principle. This was put forward by Werner Heisenberg in 1927, and states that *it is impossible to know precisely both the position and the velocity of an electron simultaneously*. If the velocity of an electron is measured accurately, a measurement of its position will be less accurate; and vice versa.

The Hydrogen Atom. We shall now consider the structure of an atom in the light of the theory of *quantum mechanics*.

According to the uncertainty principle, it is impossible to specify exactly the position of an electron of known velocity (and therefore of known energy). In discussing the motion of an electron about the nucleus, therefore, we can only say that the probability of finding that electron is greater near the nucleus than it is further away from the nucleus. So that even though the electron of a hydrogen atom may be rotating around the nucleus at varying distances from it, more often than not it will be found somewhere close to the proton.

If the atom is represented pictorially by marking the approximate position of the electron at any instant by a 'smear', then after an interval of time it might be visualised as shown on page 6. This 'picture' of the motion of an electron about the nucleus is called an *electron cloud*. The darker region near the nucleus is where the electron will spend most of its time.

It is possible to calculate the probability of finding an electron at a particular point in the space around the nucleus, and this probability is called the *electron density*. The darker regions in the above diagram represent areas of high electron density.

It can be seen that the atom does not possess a definite radius, since the electron cloud is not confined within a boundary.

The electron can no longer be described as moving in an orbit of fixed radius, and the term *atomic orbital* is used to describe the

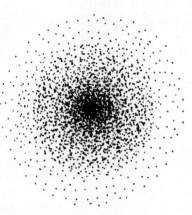

wave motion of electrons in atoms. In 1926, Erwin Schroedinger described electron orbitals in terms of mathematical functions by means of his *wave equation*. Quantum mechanics involve a considerable amount of mathematics, and it is difficult to understand many of the concepts without resorting to the use of equations. The fact that we are not making use of mathematical tools in this book will limit our appreciation of some finer points of the theory. This is mentioned not as an excuse, but as an incentive for you to look deeper into the subject.

Before we continue with the description of atomic orbitals, we may compare the structure of an atom according to the Bohr theory with that derived from quantum mechanics. Bohr considered that the electrons occupy planet-like *orbits* around the nucleus; from the quantum mechanical viewpoint, they occupy *orbitals* which cannot be completely defined.

CHAPTER 2

ATOMIC ORBITALS

As we have seen in Chapter 1, it is impossible to define the motion of an electron precisely. However, solution of the Schroedinger equation tells us that only certain orbitals and energy levels are allowed. It enables us to calculate these levels and hence to estimate the regions in space where the electron is most likely to be found. These regions are called atomic orbitals. When an electron is moving in an atomic orbital, the energy of the system does not vary with time. Such a system is known as a *stationary state*.

Specification of Orbitals by Quantum Numbers. The allowed atomic orbitals can be obtained by solution of the Schroedinger equation; although this is a complicated process, the results are very simple. It turns out that the allowed orbitals are determined by a set of numbers which are known as *quantum numbers*. These quantum numbers determine the energy of a particular orbital, and the space occupied by that orbital. Two important quantum numbers are given by n and l; and each orbital has an allowed value of n and l. The first quantum number n is called *the principal quantum number*, and it mainly determines the energy and size of the orbital. The second quantum number l modifies the energy and also determines the shape of the orbital.

The principal quantum number n can have any integral value from 1 to infinity (although values greater than 7 are seldom encountered); whereas l is restricted by the value of n. For a particular value of n, l may have values of $n-1$ and all intermediate integral values down to zero.

It is possible for some orbitals to be identical with respect to energy and shape (same n and l), but to differ only in their orientation in space. For a particular value of l, there are $2l+1$ orbitals having the same energy. For example, if $n=2$, l can have values of 0 and 1; thus there are two sets of orbitals, one having quantum numbers $n=2$, $l=0$, and the other having quantum numbers $n = 2$, $l = 1$. The first set contains only one orbital (since $2l + 1 = 1$), whilst the second contains three orbitals (since $2l + 1 = 3$).

The table below give the values of n and l, and also the maximum number of orbitals in each *principal quantum shell* (which is defined by n).

n	l	Number of orbitals for a given value of l (i.e. $2l+1$)	Total number of orbitals in shell
1	0	1	1
2	0	1	4
	1	3	
3	0	1	
	1	3	9
	2	5	
4	0	1	
	1	3	16
	2	5	
	3	7	

Note. The principal quantum shells are known as K, L, M, N, ... for $n = 1, 2, 3, 4 \ldots$

An atomic orbital is defined by using the quantum number n to specify the shell in which the orbital occurs, together with the appropriate l value.

Orbitals with l values of 0, 1, 2 and 3 are referred to respectively as s, p, d and f. So that if $n = 1$, $l = 0$ we have a $1s$ orbital Similarly, when $n = 2$ and $l = 1$, we have a $2p$ orbital.

The table below summarises the atomic orbitals which have been discussed.

Principal quantum shell (given by n)	Type of orbital (given by l)	Number of orbitals	Nomenclature of orbital
1	s	1	$1s$
2	s	1	$2s$
	p	3	$2p$
3	s	1	$3s$
	p	3	$3p$
	d	5	$3d$
4	s	1	$4s$
	p	3	$4p$
	d	5	$4d$
	f	7	$4f$

Relative Energies of Atomic Orbitals. Each set of atomic orbitals has a different energy level; and for an electron to move from one to another, energy must be either absorbed or dissipated.

The relative energies of the orbitals in atoms are shown in the diagram below.

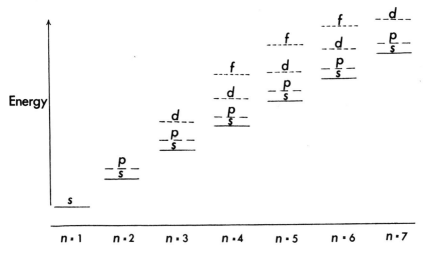

Note. Since n determines orbital size, then as n becomes greater so the size of the orbital is increased, i.e. the $3s$ orbital is larger than the $2s$, which in turn is larger than $1s$. In effect, this means that $3s$ electrons spend most of their time at distances further from the nucleus than do $2s$ electrons, and hence have a greater energy (see p. 3).

Electrons enter the orbital of lowest energy first; so that in the hydrogen atom, which has only one electron, it will normally occupy the $1s$ orbital. This is the most stable state for an atom of hydrogen and is called the *ground state*. Other atoms have more than one electron; and as each successive orbital is filled, the remaining electrons are accommodated in the orbital of next lowest energy until that also is filled, and so on. For example, from the diagram we can see that the $3s$ orbital is filled before the $3p$ orbital, but that the $4s$ is filled before the $3d$.

Accommodation of Electrons in Atomic Orbitals. We have seen that orbitals are filled in accordance with their relative energies; but we have not yet established the number of electrons which an orbital can accommodate. According to the *Pauli exclusion principle* an orbital can contain only two electrons; and these two electrons must have opposite spins (see p. 4). We can now determine the total number of electrons which can be accommodated in each shell:

Type of orbital	Number of orbitals	Total number of electrons that can be accommodated	
		In set	In shell
$1s$	1	2	2
$2s$	1	2	8
$2p$	3	6	
$3s$	1	2	
$3p$	3	6	18
$3d$	5	10	
$4s$	1	2	
$4p$	3	6	32
$4d$	5	10	
$4f$	7	14	

Let us illustrate the application of the atomic orbital theory to atomic structure by working out the structures of the first few elements in the periodic table.

Hydrogen. We have already seen that hydrogen (atomic number 1) has its electron in the $1s$ orbital. The electronic configuration can be written as $1s^1$.

Helium (Atomic Number 2). Both electrons are in the $1s$ orbital which is therefore completely filled, as is the first principal quantum shell, i.e. the electronic configuration is $1s^2$. Each electron has opposite spin (see pp. 4 & 10).

Lithium (Atomic Number 3). There are three electrons; two of these are in the $1s$ orbital and the other goes into the orbital of next lowest energy, $2s$.

We can denote the electronic configuration of lithium like this:

$$\boxed{|\,|}\quad\boxed{|}\quad 1s^2 2s^1$$
$$\;\;1s\quad\;\;2s$$

where each box represents an atomic orbital, and each vertical line therein represents an electron in that orbital.

Beryllium (Atomic Number 4). The electronic configuration is:

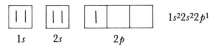

Boron (Atomic Number 5). Since the $2s$ orbital is filled, one electron must go into the orbital of next lowest energy, i.e. $2p$.

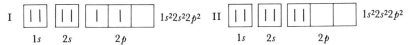

Since there are three $2p$ orbitals they are represented by three adjacent boxes.

Carbon (Atomic Number 6). In this case there are apparently two alternatives:

I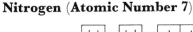

To determine which of these two structures is correct we need to use a theory put forward by Hund; and for our purposes we can modify this theory and state that, in filling orbitals of equal energy, the most stable configuration is when the electrons are 'unpaired', i.e. *when the two electrons are in different orbitals.*

Structure I therefore, is the most stable for the carbon atom.

Structure II has the higher energy partly because the two electrons are in the same region of space (orbital), and will therefore repel each other more strongly than if they were in separate orbitals.

Nitrogen (Atomic Number 7)

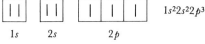

Oxygen (Atomic Number 8)

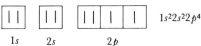

Fluorine (Atomic Number 9)

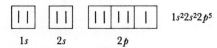

$1s^2 2s^2 2p^5$

1s 2s 2p

Neon (Atomic Number 10)

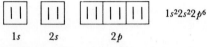

$1s^2 2s^2 2p^6$

1s 2s 2p

The Periodic Table. The periodic table of the elements is of paramount importance in the study of chemistry, and it is well worth taking a look at it in the light of the atomic orbital theory.

Study it carefully; you should be able to understand its meaning from what we have already said about electronic orbitals.

The atomic number of each element is given above its symbol in the periodic table; and it tells us how many electrons the atom possesses.

Group I consists of elements with one s electron in their outer shells. If we work out the order in which the orbitals are filled with electrons for the elements of Group I we should be able to see immediately what the structure of all the other elements will be.

Element	H	Li	Na	K	Rb	Cs	Fr
Atomic number	1	3	11	19	37	55	87
Order in which the orbitals are filled	$1s^1$	$1s^2$ $2s^1$	$1s^2$ $2s^2$ $2p^6$ $3s^1$	$1s^2$ $2s^2$ $2p^6$ $3s^2$ $3p^6$ $4s^1$	$1s^2$ $2s^2$ $2p^6$ $3s^2$ $3p^6$ † $\begin{cases}4s^2\\3d^{10}\end{cases}$ $4p^6$ $5s^1$	$1s^2$ $2s^2$ $2p^6$ $3s^2$ $3p^6$ $4s^2$ $3d^{10}$ $4p^6$ $5s^2$ $4d^{10}$ $5p^6$ $6s^1$	$1s^2$ $2s^2$ $2p^6$ $3s^2$ $3p^6$ $4s^2$ $3d^{10}$ $4p^6$ $5s^2$ $4d^{10}$ $5p^6$ $6s^2$ $4f^{14}$ $5d^{10}$ $6p^6$ $7s^1$

† This has already been remarked upon (p. 9). Use the relative orbital energy diagram on p. 9 together with the table on p. 10 when working out the order in which the orbitals are filled.

[Ch. 2] *Atomic Orbitals* 13

PERIODIC TABLE OF THE ELEMENTS

Period		Gp I	Gp II					Transition metals						Gp III	Gp IV	Gp V	Gp VI	Gp VII	Gp 0
1	$1s$	1 H																	2 He
2	$2s$ $2p$	3 Li	4 Be											5 B	6 C	7 N	8 O	9 F	10 Ne
3	$3s$ $3p$	11 Na	12 Mg											13 Al	14 Si	15 P	16 S	17 Cl	18 Ar
4	$4s$ $3d$ $4p$	19 K	20 Ca	21 Sc	22 Ti	23 V	24 Cr	25 Mn	26 Fe	27 Co	28 Ni	29 Cu	30 Zn	31 Ga	32 Ge	33 As	34 Se	35 Br	36 Kr
5	$5s$ $4d$ $5p$	37 Rb	38 Sr	39 Y	40 Zr	41 Nb	42 Mo	43 Tc	44 Ru	45 Rh	46 Pd	47 Ag	48 Cd	49 In	50 Sn	51 Sb	52 Te	53 I	54 Xe
6	$6s$ $4f^{*}$ $5d$	55 Cs	56 Ba	57 La*	72 Hf	73 Ta	74 W	75 Re	76 Os	77 Ir	78 Pt	79 Au	80 Hg	81 Tl	82 Pb	83 Bi	84 Po	85 At	86 Rn
7	$7s$ $5f\dagger$ $6d$	87 Fr	88 Ra	89 Ac†															

*Lanthanide series $4f$	58 Ce	59 Pr	60 Nd	61 Pm	62 Sm	63 Eu	64 Gd	65 Tb	66 Dy	67 Ho	68 Er	69 Tm	70 Yb	71 Lu
†Actinide series $5f$	90 Th	91 Pa	92 U	93 Np	94 Pu	95 Am	96 Cm	97 Bk	98 Cf	99 Es	100 Fm	101 Md	102 No	103 Lw	104

Note. The orbitals given below each period number show the order in which they are filled in that period. Gp = Group.

Note that the electronic configuration is written in the order of increasing n, and not in the order in which the orbitals are filled. For example, the electronic configuration of Rb is

$$1s^2 2s^2 2p^6 3s^2 3p^6 3d^{10} 4s^2 4p^6 5s^1.$$

Each period in the table takes us from a Group I element to a noble gas (Group **0**). These have outer shells of $s^2 p^6$ electrons (except He, which is $1s^2$); so that the following period starts again with one electron in the next shell.

The *transition metal* atoms are built up by filling d orbitals; and the *lanthanide series* by filling $4f$ orbitals.

Group III elements have three electrons in their outer shells (two s and one p). Similarly, Group IV elements have four electrons in their outer shells; Groups V, VI and VII have five, six and seven electrons respectively in their outer shells.

The *actinide series* is built up by filling $5f$ orbitals.

Beneath the valence (outer) shell there is a noble gas 'core' for all elements of Groups I and II; whilst for the other groups (apart from the first two elements of the group, which have the noble gas core) the remaining elements have either ten or fourteen electrons directly beneath the outer (or valence) shell. The elements of Groups **0** to VII are known as *main group elements*.

The configurations of the underlying shells are important because these are the ones which 'screen' the valence electrons from the positive nuclear charge. They have a bearing on the 'size' of the atom; and the ease with which an atom may lose an electron. Thus for Groups I and II elements, which each have a noble gas core, there is a fairly regular decrease in ionisation potential (see p. 18) and increase in atomic size with increasing atomic number. For other groups, the underlying shells differ in configuration from one another and we may therefore expect less regular trends.

We now have a fairly complete picture of the atomic structures of all the elements, and we can proceed to discuss how these structures influence the way in which atoms combine with one another.

CHAPTER 3

BASIC PRINCIPLES OF VALENCE THEORY

Why do some atoms combine together and others do not? When they do combine, why do they do so in certain definite proportions and not in others? What determines the shape of molecules? These are some of the questions which a theory of valency must answer.

Chemical experiments show that there are roughly two types of bond formed when atoms unite to give molecules; *ionic* and *covalent*. We shall see, however, that when bonding of atoms is considered in the light of quantum mechanics, the terms 'ionic' and 'covalent' describe only the extreme types of bond; real ones are intermediate between the two extremes.

Let us consider what happens as two neutral atoms approach each other. When the atoms are far enough apart for their electron clouds to be completely separate from one another, there is no force between them. As the atoms move nearer together, the electron clouds overlap and there is an attraction between the electrons in one atom and the nucleus of the other. There is also repulsion between the electrons from each atom and repulsion between the two nuclei. These repulsive forces become dominant at small distances.

We can illustrate this by means of the energy curve on page 16. This is a plot of the sum of the electronic energy and nuclear repulsion energy of a stable diatomic molecule against the internuclear distance.

Overlapping of the atomic orbitals leads to a lowering of the energy, due to the attractive forces between the electrons and the nuclei. The energy continues to fall until the nuclei are so close together that repulsion occurs, and the energy increases. The energy minimum is when the atoms are said to be bound together; a chemical bond has been formed. The internuclear

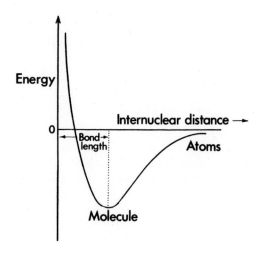

distance at the position of minimum energy is the *bond length*, but this is not rigidly fixed (remember the uncertainty principle) —it is only the *mean nuclear separation*.

Atoms combine because their total energy is lowered by combination with each other.

We shall now consider the extreme types of bond.

The Ionic (or Electrovalent) Bond. An ionic bond is formed by an atom losing one or more electrons to another atom in order to attain a more stable configuration (and thereby lowering the total energy of the system).

We must first determine what is a stable configuration. The elements of Group **0** are noted for their poor chemical reactivity (until recently they were not known to react with anything, and were hence called the 'inert' gases). Helium has the configuration $1s^2$; it has a completely filled shell of electrons. The other members of Group **0**, namely neon, argon, krypton, xenon and radon, have the electronic configuration ns^2np^6 in their outer shells; i.e. the s and p orbitals are completely filled. This configuration is extremely stable and can be achieved by other atoms, for example sodium and fluorine when they form sodium fluoride:

$$\text{Na} \quad 1s^22s^22p^63s^1$$
$$\text{F} \quad 1s^22s^22p^5$$

If the 3s electron is transferred from the sodium atom to the fluorine atom, both will acquire the configuration of neon:

$$\text{Ne} \quad 1s^2 2s^2 2p^6$$

We can write this as:

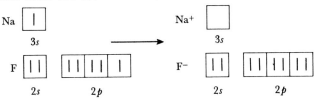

Note. It is only necessary to consider electrons in the outer (or valency) shell.

The resultant molecule of sodium fluoride consists of a positively charged sodium ion (cation) and a negatively charged fluoride ion (anion).

More than one electron may be transferred, as in calcium chloride:

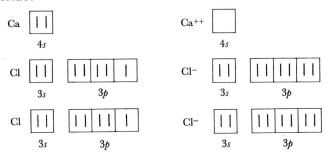

Although many of the ions in ionic compounds possess the electronic configurations of Group **0** elements, this is not always the case. Other factors are also important, and we shall now discuss the criteria which lead to conditions that are energetically favourable to the formation of ionic bonds.

Ionisation Potential. The ionisation potential I is the energy required to remove an electron from an atom in the gaseous state:

$$\text{atom} \xrightarrow{+I} \text{cation} + e^-$$

The lower the value of I, the more chance there is of an ionic bond being formed.

Since all atoms (with the exception of hydrogen) possess more than one electron, there are several ionisation potentials for the same atom; only those of the valency electrons need to be considered. For example, for a divalent atom M the first ionisation potential I_1 is for the removal of the first valency electron, i.e.:

$$M \xrightarrow{I_1} M^+ + e^-$$

The second ionisation potential, I_2, is for the removal of an electron from the M^+ ion, i.e.:

$$M^+ \xrightarrow{I_2} M^{2+} + e^-$$

The first ionisation potentials (i.e. I for the removal of the first valency electron, as above) of the elements vary with their position in the periodic table. In the groups, I usually decreases with increase in atomic number. In the periods, I increases with atomic number.

The values for I in kcal/mol for the first four groups and the second and third periods are given below.

Increase of I →

Decrease of I ↓

Li	124	Be	215	B	191	C	260	N	335	O	314	F	402	Ne	497
Na	118	Mg	176	Al	138	Si	188	P	254	S	240	Cl	300	Ar	363
K	100	Ca	140	Ga	138	Ge	187								
Rb	96	Sr	131	In	134	Sn	168								
Cs	90	Ba	120	Tl	141	Pb	171								

Note. 1 kcal/mol = 4·184 kJ/mol.

For the first two groups there is a decrease in I as the average distance of the electron from the nucleus increases; whereas in the other groups there are irregularities in this trend (see p. 14).

We shall refer again to ionisation potentials after considering electron affinity.

Electron Affinity. The electron affinity A is the energy released by an atom in the gaseous state when an electron is added to it (i.e. the electron affinity is the ionisation potential of the negative ion):

$$\text{atom} + e^- \xrightarrow{-A} \text{anion}$$

The higher the value of A the more chance there is of an ionic bond being formed.

Group VII (the halogens) have the highest electron affinities; this may be attributed to the completion of Group **0** (the noble gases) configurations when they gain an electron.

Group I (the alkali metals) have very low electron affinities since their outermost electron is not bound very strongly.

The electron affinities of Group VII are given below, and compared with that of sodium.

	kcal/mol	kJ/mol
F	84	351
Cl	87	364
Br	82	343
I	75	314
Na	17	71

From what has been said of ionisation potentials and electron affinities, it is evident that the most stable ionic bonds will be formed between atoms with low values of I combining with atoms of high values of A. The atom with the lowest value of I is caesium (90 kcal/mol), and that with the highest value of A is chlorine (87 kcal/mol). These two atoms will therefore form a very stable ionic molecule. From energy considerations, however, we find that there would appear to be an *increase* in energy when CsCl is formed!

$$Cs \xrightarrow[367 \text{ kJ}]{90 \text{ kcal}} Cs^+ + e^-$$

i.e. * absorption of 90 kcal is required to bring about ionisation.

$$Cl + e^- \xrightarrow[-364 \text{ kJ}]{-87 \text{ kcal}} Cl^-$$

i.e. * liberation of 87 kcal occurs when an electron is added to the chlorine atom.

Thus there is resulting absorption of energy, i.e. a net increase in energy, of about 3 kcal (12kJ), for the reaction

$$Cs + Cl \longrightarrow Cs^+ + Cl^-$$

If we also take into account the quantities of energy required to form gaseous caesium atoms from solid metal and chlorine atoms from chlorine molecules, both of which are positive, the energy increase is considerably greater. This seems peculiar! We have said that atoms combine in order to decrease their total

---

* Sign convention: if energy is *liberated*, the sign is *negative*; if energy is *absorbed*, the sign is *positive*.

energy; and from considerations of ionisation potential and electron affinity it is evident that CsCl should possess the most stable ionic bond. Yet energetically that would appear to be impossible, and on this basis we should expect that no ionic bonds could be formed! Clearly, we have not yet taken everything into consideration. Since ionic compounds exist in the crystalline state, and so far we have considered only isolated ions in the gaseous state, we need also to consider the energy changes involved in forming a crystal from gaseous ions.

Lattice Energy. The lattice energy is the energy released when gaseous ions form one gramme-molecule of compound in a crystal lattice. It results in the system losing energy, owing to electrostatic attraction between positive and negative ions. *A high value for the lattice energy will favour the formation of an ionic bond.*

Ionic binding only occurs if the electrostatic attraction of the ions can outweigh the energy difference between the ionisation potential I and the electron affinity A. The lattice energy of CsCl is approximately 150 kcal/mol (628 kJ/mol).

We may now summarise some of the conditions which favour the formation of ionic bonds:

1. The atoms may easily acquire a noble gas electronic configuration when they become ions.
2. A low ionisation potential (this refers to the metal atom).
3. A high value for the electron affinity (this refers to the non-metallic atom).
4. A high value for the lattice energy of the ionic compound which is formed.

Most ionic compounds are formed by the combination of elements from opposite sides of the periodic table, i.e. halides, sulphides and oxides of Groups I and II (and also of the transition metals).

Crystal Structure. A characteristic feature of ionic compounds is that they occur in aggregates of ions which form a crystal structure; and from the arrangement of ions in the crystal it is possible to calculate the lattice energy.

The manner in which ions are packed into a crystal lattice depends much on the relative sizes of the anion and the cation.

[Ch. 3] Basic Principles of Valence Theory

By regarding the ions as hard spheres and considering their relative sizes, it is possible to predict the type of crystal lattice which will be formed.

We shall now consider how the constituent ions group together to form crystals.

The important point about the structure of crystals is that they are formed by the packing together of small repeating arrangements of ions, which are known as *unit cells*. The crystal lattice so formed consists of positively charged ions (cations) surrounded by negatively charged ions (anions), and vice versa.

A diagram of the unit cell of caesium chloride is shown below.

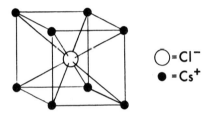

Note. Lines are drawn joining the ions in order to clarify the three-dimensional arrangement.

This type of unit cell is termed *body-centred cubic*.

When the crystal lattice is built up from the unit cells, there is one pair of ions per unit cell. (The unit cell as shown in the diagram could not exist on its own; we get a true picture of the crystal structure only when we see a unit cell in relation to other unit cells.)

It can be seen from the diagram of a crystal lattice on page 22 that each cation (Cs^+) is surrounded by eight anions (Cl^-) at equal distances. Similarly, each Cl^- is surrounded by eight Cs^+ (see diagram above). The overall ratio of Cs^+ to Cl^- is therefore one to one.

This type of crystal lattice is exhibited by several ionic substances, including the chlorides, bromides and iodides of caesium and thallium.

A more common structure (*face-centred cubic*) is exhibited by almost all the alkali halides and the oxides and sulphides of the

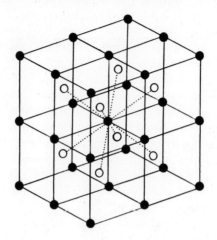

alkaline earth metals (Group II). The structure of sodium chloride is shown below.

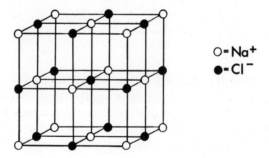

○ = Na⁺
● = Cl⁻

The cations (Na⁺) are at the corners and centres of the faces of the cube (hence the name face-centred cubic). The anions (Cl⁻) also have the same arrangement. Each cation is surrounded by six anions, and vice versa, to give an overall ratio of 1:1 for $Na^+ : Cl^-$.

If you make simple models to illustrate these arrangements of ions, you will understand their structures more readily.

Whilst in many cases the predicted crystal lattices are found, there are compounds which deviate from the expected structure. This is often due to the bonds not being completely ionic.

Ionic bonds, unlike covalent bonds as we shall see, are non-directional.

The Covalent Bond. We have seen how an ionic bond is formed by a metal atom losing one or more loosely bound outer electrons to a non-metallic atom (or atoms); the resulting ions then fall together into a crystal lattice. Covalent bonds, on the other hand, are produced by the *sharing* of electrons between atoms to form molecules. In this case, ions are not involved.

The energy changes which occur when two neutral atoms approach one another have already been discussed (p. 15). Overlap of atomic orbitals from the two atoms causes attractive forces between the electrons from one atom and the nucleus of the other. As the atomic nuclei come closer together there is repulsion, but this is moderated to some extent by the intervening electrons, and a position of equilibrium between attractive and repulsive forces is reached. At this point the energy is at a minimum, and a covalent bond has been formed. Covalent bonding may be considered as a *pairing of electrons* by the overlap of two atomic orbitals from different atoms.

In order to illustrate the manner in which electrons are shared between atoms, we shall consider the electronic configurations of elements in Period 2 of the periodic table: beryllium, boron, carbon, nitrogen, oxygen and fluorine. The order in which we look at these elements will be one of convenience.

Nitrogen. For the pairing of electrons by overlapping atomic orbitals, we need a single unpaired electron in each orbital that is involved in the formation of a bond.

The electronic configuration of nitrogen is $1s^2 2s^2 2p^3$, and the valence electrons can be represented as

$\quad\quad\quad\quad 2s \quad\quad\quad 2p$

There are three unpaired electrons in the outer shell, and these are available for the formation of covalent bonds. With H for example, which has the configuration

$\quad\quad\quad 1s$

ammonia will be formed.

This can be represented as follows:

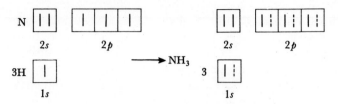

Note. The broken vertical lines represent electrons which are shared between two orbitals.

We have already described how the completion of ns^2np^6 leads to a stable configuration (p. 16).

Nitrogen is said to have a *covalency* of three.

You will notice that there are two electrons in the $2s$ orbital which are not being used; these are known as a *lone pair* of electrons. We shall be discussing these later.

Oxygen. O has the electronic configuration $1s^2 2s^2 2p^4$, i.e.:

$$\boxed{|\,|}\quad \boxed{|\,|\;|\;|}$$
$$\;2s\qquad\quad 2p$$

There are two unpaired electrons, and these can be used to form the water molecule for example:

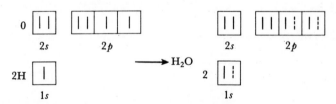

Oxygen therefore, has a covalency of two.

Fluorine

$1s^2 2s^2 2p^5$, $\boxed{|\,|}\quad \boxed{|\,|\;|\,|\;|}$
$\qquad\qquad\quad 2s \qquad\;\; 2p$

Only one orbital contains an unpaired electron; fluorine therefore, has a covalency of one, e.g.:

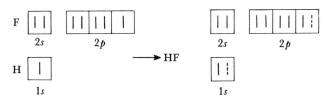

Beryllium

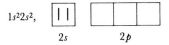

Beryllium has no unpaired electrons, and would be expected to have zero covalency. It shows some similarity to helium ($1s^2$), which also has a completely filled s orbital. There is a difference however, since the $2n$ shell possesses p orbitals and the $1n$ shell does not. Beryllium can be made to produce unpaired electrons by promoting one of the $2s$ electrons into the p orbital. To do this, it is necessary to add energy to the system; this energy is called the *promotion energy*, i.e.

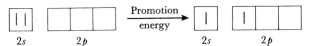

When this happens, it becomes possible to overlap the two orbitals with those from another atom. A stable molecule will be formed only if the net energy released by the formation of the two bonds is greater than the energy required to promote an electron from the $2s$ orbital to the $2p$ orbital. This occurs in the formation of beryllium chloride ($BeCl_2$), in which beryllium shows a covalency of two:

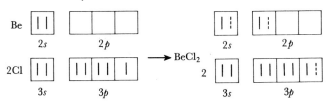

For helium, the energy involved in promoting a 1s electron to the 2s orbital is so great that any energy released by the formation of bonds is prohibited; no compounds are formed therefore. In general, the energy changes are so great as to make promotion of an electron from one principal quantum shell to another impossible in chemical reactions.

Carbon

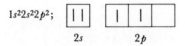

From the electronic configuration, carbon would be expected to form divalent compounds. These are not usually found, however, and a covalency of four predominates. This can be achieved by the promotion of a 2s electron into the p orbital:

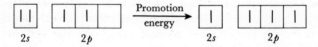

There is a great deal of energy liberated by the formation of four bonds, and the promotion energy is easily overcome. For example:

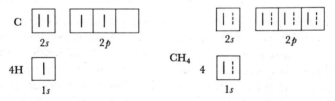

Note. The energy released in the formation of four covalent bonds is far greater than in the formation of a divalent compound.

Boron

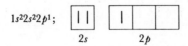

The same considerations apply to boron as to carbon. Boron therefore, forms stable trivalent compounds.

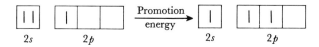

For example:

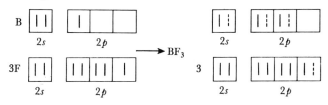

Note that in trivalent boron compounds there is an empty p orbital; and in simple beryllium compounds there are two empty p orbitals. All the other atoms we have considered so far have achieved a stable 'octet' of electrons when they form covalent compounds. From this we can conclude that although the formation of an electronic configuration similar to that of the noble gases (i.e. an octet of electrons in the valence shell) leads to stability, it is by no means a requirement for the formation of stable covalent compounds. The presence of vacant orbitals in the valence shell, however, affects the chemical properties considerably. These orbitals may overlap those containing a 'lone pair' of electrons from other molecules (ammonia for example) to form compounds. In this case, one atom is providing *both* the electrons, and the bond formed is called a *coordinate* (or dative covalent) bond. Once a coordinate bond has been formed, there is no difference between it and a normal covalent bond.

As an example of coordinate bonding, we can take the compound formed between ammonia and boron trifluoride:

$$NH_3 + BF_3 \longrightarrow H_3N \rightarrow B {\begin{array}{c} F \\ F \\ F \end{array}}$$

The vacant p orbital of BF_3 overlaps the lone pair in the NH_3 molecule.

We may now consider some of the elements in Period 3 of the periodic table. Here, we have atoms with valence electrons in the third principal quantum shell $(n = 3)$; and in addition to s and p orbitals, there are also the $3d$ orbitals to consider. In view of this, we can expect a wider range of covalencies to be exhibited by these elements.

Phosphorus. The electronic configuration of P is:
$$1s^2 2s^2 2p^6 3s^2 3p^3.$$
Considering the valence electrons only, this can be written:

It can be seen that phosphorus will exhibit a covalency of three, as in phosphorus trifluoride:

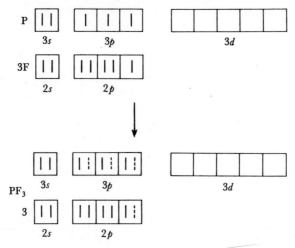

Phosphorus can also show a covalency of five, by the promotion of a $3s$ electron into a $3d$ orbital:

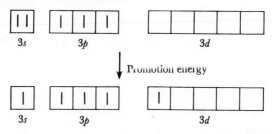

For example, in the formation of phosphorus pentafluoride PF$_5$:

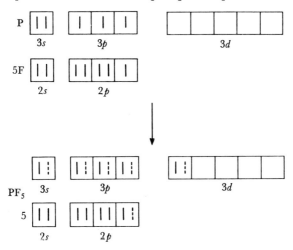

Phosphorus, therefore, can exhibit covalencies of three and five.

Sulphur

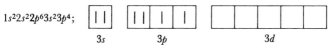

Sulphur exhibits a covalency of two, as in hydrogen sulphide:

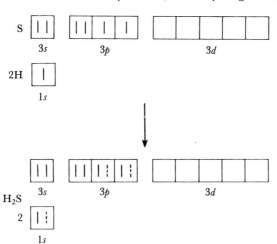

It will also show a covalency of four, by promotion of a $3p$ electron to the $3d$ orbital:

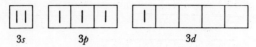

For example, in the formation of sulphur tetrafluoride (SF_4):

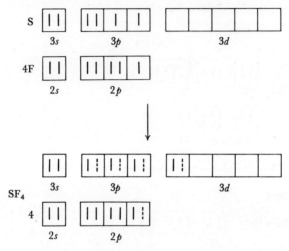

A covalency of six can be exhibited by promoting both a $3s$ and a $3p$ electron into the $3d$ orbital:

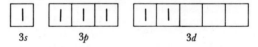

For example, sulphur hexafluoride (SF_6):

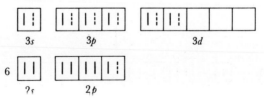

Thus, sulphur can have covalencies of two, four and six.
You will now see how it is possible for all the valence electrons to take part in covalent bonding; and that the atomic orbital

theory explains very nicely the multiple covalency exhibited by some elements.

Chlorine can have covalencies of one, as in HCl; three, as in chlorine trifluoride ClF_3; and five, as in chlorine pentafluoride ClF_5. In principle, chlorine should also be able to exhibit a covalency of seven (to form a compound such as ClF_7), but this does not occur, perhaps owing to the difficulty of packing seven fluorine atoms round one atom of chlorine.

It may be mentioned, however, that iodine (which is a larger atom than chlorine) does form a compound (IF_7) in which it has a covalency of seven; this is one of the few known compounds in which this valency is encountered. Iodine also has covalencies of one, three and five; as in IF, IF_3 and IF_5.

Shapes of Molecules. It is possible to predict, using very simple ideas, the shapes of most covalent molecules formed by the main group elements.

The shapes of molecules are determined by their electronic configurations. Elements which, when they form molecules, have the same configurations of electrons in their valence shells, give the same shaped molecules. *Shape is dependent upon the number of electron pairs in the valence shell*; this is because each electron pair is repelled by the others, and they therefore orientate themselves in space so as to be as far away from each other as possible. We shall now look at a few simple molecules to illustrate this.

Beryllium Chloride ($BeCl_2$). The electronic configuration of Be in $BeCl_2$ can be represented as:

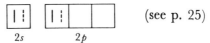 (see p. 25)

There are *two electron pairs* in the valence shell; and if these are to be as far away from one another as possible they must make an angle of 180° at the nucleus. The $BeCl_2$ molecule is therefore *linear*:

Cl—Be—Cl

Note. $BeCl_2$ is only monomeric (i.e. exists as separate molecules) in the gaseous state.

Boron Trifluoride (BF₃). The electronic configuration of B in BF_3 can be represented as:

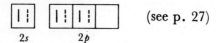

There are *three electron pairs* in the valence shell; and for them to be as far away from each other as possible, the molecule must be *triangular planar*:

Methane (CH₄). In CH_4, the carbon atom has acquired *four electron pairs*:

(see p. 26)

Methane is *not* square planar (i.e. H—C—H with H above and below), since this would mean that the electron pairs would make an angle of only 90° at the nucleus. In a *tetrahedral* arrangement, the angle is 109° 28′; the electron pairs are therefore farther apart than they would be in a square planar arrangement.

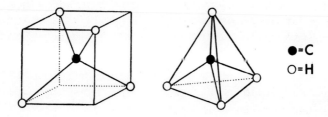

● = C
○ = H

[Ch. 3] *Basic Principles of Valence Theory* 33

The atoms of H are positioned at the corners of a regular tetrahedron with C in the centre. It can be seen from the above diagram that the hydrogen atoms occupy four corners of a cube.

Ammonia (NH₃). In NH_3, the N atom can be represented as:

There are three electron pairs involved in bonding, and also a lone pair. These four pairs of electrons are arranged tetrahedrally around the N nucleus:

The molecule of NH_3 is pyramidal:

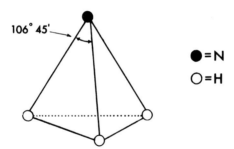

● = N
○ = H

The angle formed between two atoms of H and the N atom (i.e. the *bond angle*) is 106° 45′. This is because the lone pair of electrons has a greater repulsive force* than has an electron pair involved in bonding (it occupies more space since it is not restricted to two atoms); otherwise the bond angle would be 109° 28′ as in a regular tetrahedron.

The ammonium ion, however, forms a regular tetrahedron

---

* Repulsion between pairs of electrons decreases in the order: lone pair − lone pair > lone pair − bonded pair > bonded pair − bonded pair.

(bond angle = 109° 28'). When the lone pair is involved in bonding, the repulsion is the same as for an electron pair.

Water (H_2O). In H_2O, the O atom has four pairs of electrons; two of these are involved in bonding, and there are two lone pairs:

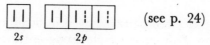   (see p. 24)

Thus the water molecule is:

The bond angle is 104° 27', owing to repulsion from the two lone pairs of electrons.

Phosphorus Pentafluoride (PF_5). In this case there are *five electron pairs*:

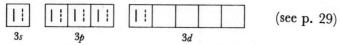   (see p. 29)

The shape of the PF_5 molecule is *trigonal bipyramidal*:

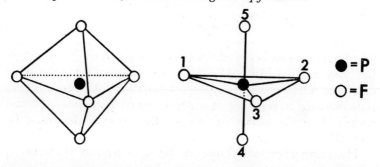

Three of the fluorine atoms are at the corners of an equilateral triangle, the other two are above and below the plane of this triangle as shown. The bond angles between fluorine atoms 4 and 5 and each of the others are 90°. The angles between fluorine atoms 1, 2 and 3 are 120°.

Sulphur Tetrafluoride (SF$_4$) In SF$_4$ there are four electron pairs involved in bonding and one lone pair:

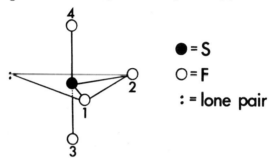

The arrangement therefore, will be trigonal bipyramidal:

The bond angles are: between fluorine atoms 3 and 4, 173°; and between fluorine atoms 1 and 2, 101°.

Note. It may be thought that the lone pair could alternatively be in position 4; it can be shown, however, that this is not so. A useful rule to remember is that the most electronegative elements (see p. 37) occupy the axial positions (3 and 4); these will therefore be occupied by F atoms. Also, for compounds such as PF$_3$Cl$_2$, the structure will be:

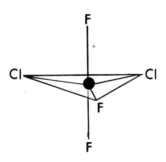

Sulphur Hexafluoride (SF_6). There are *six electron pairs* in SF_6:

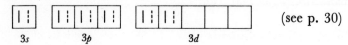

(see p. 30)

The shape of the molecule is that of a *regular octahedron*:

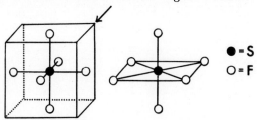

Viewed from the arrow point, the octahedral arrangement of atoms would look like this:

The bond angles are each 90°.

Iodine Pentafluoride (IF_5). There are five pairs of bonding electrons and one lone pair, i.e. six pairs of electrons in the valency shell of I in IF_5:

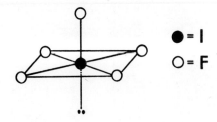

The arrangement is therefore octahedral:

The bond angles have not yet been measured accurately, but we should expect them to be less than 90° owing to repulsion from the lone pair.

Iodine Heptafluoride (IF_7). In this case there are *seven pairs of bonding electrons*:

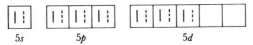

The shape of the molecule is *pentagonal bipyramidal*:

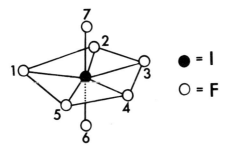

Five of the fluorine atoms are at the corners of a regular pentagon; the other two are above and below the plane of the pentagon as shown.

The bond angles between fluorine atoms 1, 2, 3, 4 and 5 are each 72°; and between 6 and 1 etc., 90°.

In summary, we can list the shapes of molecules according to the number of electron pairs in the valence shell of the central atom:

Number of pairs of electrons	Shape of molecule
2	Linear
3	Triangular planar
4	Tetrahedral
5	Trigonal bipyramidal
6	Octahedral
7	Pentagonal bipyramidal

Electronegativity. It has already been pointed out that ionic and covalent bonds are the extreme types of chemical linkage; in many compounds the bonding is intermediate between the two. In order to appreciate this, we must introduce the concept of electronegativity, which relates to the ability of an atom to attract electrons and thereby form chemical bonds.

Electronegativity can be described in terms of the ionisation potential and the electron affinity of an atom:

$$\text{Electronegativity} \propto I + A$$

If we consider a diatomic molecule AB, the ionic character of the bond can be said to be the difference in the electronegativities of A and B. The most electronegative group of elements are the halogens (Group VII), and the least electronegative elements are the alkali metals (Group I); so that the bonds formed between them will be highly ionic. Elements on the right-hand side of the periodic table have the highest electronegativities, whereas those on the left-hand side have the lowest values. A few examples are given below:

Element	Electronegativity
F	4·0
O	3·5
N	3·0
Cl	3·0
Br	2·8
S	2·5
Ca	1·0
Ba	0·9
Na	0·9
K	0·8

Compounds formed between elements from opposite sides of the periodic table will therefore possess ionic bonds (see p. 20).

Ideally, a covalent bond will be formed between atoms of the same electronegativity; this rarely occurs however, since there is usually some difference in the electronegativities of combining atoms. Thus, most covalent bonds can be said to have some ionic character (i.e. the more electronegative atom has a greater share of the electrons).

We have already fulfilled the requirements of a valence theory on the basis of atomic orbitals. We know that atoms combine together because their total energy is lowered by doing so; we know why they combine in certain proportions; and finally, we have shown how the shape of molecules is dependent upon electronic configuration. So far, however, we have given little attention to the manner in which the orbitals overlap to form a chemical bond, and we shall now discuss this in greater detail.

CHAPTER 4

THE VALENCE BOND THEORY

Chemistry has always been regarded as a practical science, and work in the laboratory tells us many things about how atoms combine to form molecules; it is often some time before a theory is found to explain experimental observations. The valence bond theory not only shows the manner in which known compounds are formed; it also enables the chemist to predict which atoms will react together and what the shapes of the resultant molecules will be. Essentially, the theory deals with the overlap of atomic orbitals—such as we have been discussing in the previous chapter; but in order for us to fully understand its significance, we must first know a little more about these orbitals.

Atomic orbitals can only be described adequately by means of mathematical functions, and it is impossible to visualise them in any other way; we cannot draw mental pictures of them. Having said this, we shall now proceed to explain how it is possible to represent graphically certain properties of orbitals, which enable us to see the manner in which they overlap.

From the *uncertainty principle* you will remember that it is impossible to specify the position of an electron; we can only say that there is a *probability* of finding it in a certain region of space. There are certain mathematical devices which can be used, however, to describe the properties of atomic orbitals. One of these refers to the probable distance of the electron from the nucleus, and another to the probable direction in which the electron may be found. It is the second mathematical function which mainly concerns us over the bonding of atoms to form molecules.

Graphical Representation of the Directional Properties of Atomic Orbitals. For convenience, we shall consider the p orbitals first.

Let us begin by simply plotting out the values that $\cos \theta$ may have for all values of the angle θ between 0° and 360°:

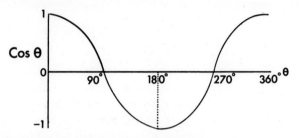

This can be plotted out in another way:
Let θ be the angle that a line OP makes with the z axis:

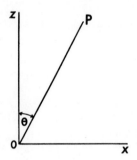

Now if we draw the length of the line OP equal to the value of $\cos \theta$, then for each value of θ we shall have a line of different length in a different direction. For values of θ between 0° and 360° we obtain two tangential circles:

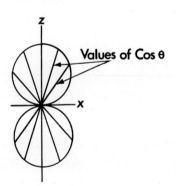

[Ch. 4] *The Valence Bond Theory* 41

θ is not restricted to two dimensions, however; and the three-dimensional plot is two spheres, one sitting upon the other:

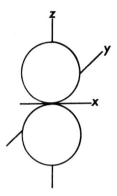

The advantage of this plot is that we can readily see how the value of cos θ changes with direction. *The value of cos θ in a particular direction of θ is obtained by measuring the length of the line from the origin to the surface of the sphere.*

It can be seen that cos θ has its maximum value along the z axis, and is zero in the xy plane.

The probability of finding an electron in a particular direction is called the directional probability or the *angular probability*. If we plot out the angular probability for an electron in a p orbital in a manner similar to that described for cos θ, then very similar diagrams are obtained. There are three p orbitals, and their angular probability diagrams are shown below.

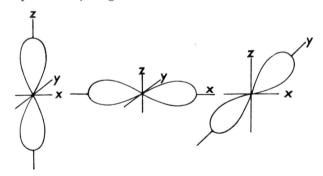

In this case, we see that the angular probability of finding an electron in a p orbital is described by a dumb-bell shaped figure;

and that the three angular probability diagrams are mutually perpendicular.

The value of the angular probability (i.e. the probability that the electron is in a given direction) for a *p* electron *is proportional to the length of the line drawn from the origin to the surface of the dumbbell*. Since there are angular probability maxima along the respective x, y and z axes, the three *p* orbitals are labelled p_x, p_y and p_z.

For simplicity, cross-sections only of the angular probability diagrams are used to show the directional properties of *p* orbitals:

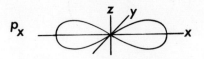

The angular probability is maximum along the x axis.

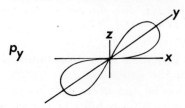

The angular probability is maximum along the y axis.

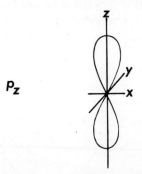

The angular probability is maximum along the z axis.

The *s* orbital. An electron in an *s* orbital may be found in *any direction* with respect to the nucleus. The probability of finding an *s* electron is the same in all directions, so that if we construct a diagram such as those for the *p* orbitals the resultant shape is a sphere:

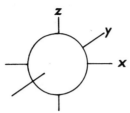

or, taking a cross-section through the sphere, we have:

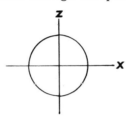

The nucleus can be considered to be at the centre of the sphere, but note that *the distance between nucleus and surface bears no relationship to the actual distance between the electron and the nucleus.* We are considering only the *directional* properties of the orbital; the probability of finding an electron at any *distance* from the nucleus is described by quite a different mathematical function, and for the 1s orbital this may be represented as follows:

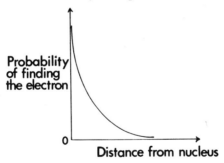

Note that there is a high probability of finding the electron close to the nucleus, and also some probability of finding it at large distances from the nucleus.

It is difficult to combine pictorially both the directional and the radial (distance) properties of an atomic orbital, and no attempt will be made here to do so. It is sufficient for our purposes to regard only the probable direction of valency electrons with respect to the nucleus.

Note. Unlike p orbitals, the s orbital has no specific directional properties; this will be found to be of significance in later discussions.

The d Orbitals. There are five d orbitals, and their angular probabilities can be represented in cross-section as follows:

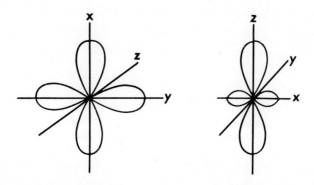

The 'bow tie' indicates that there is a small probability of finding the electron in the xy plane.

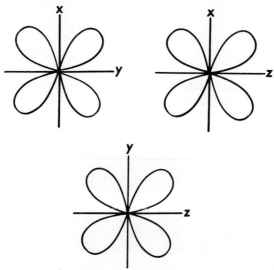

From these representations it can be seen that there are two sets of d orbitals; one set consisting of two orbitals with their maximum angular probabilities *along* the axes, and the other set of three orbitals which have their maximum angular probabilities *between* the axes. Again, directional properties are exhibited and this is of great importance for transition metal compounds.

Electron Density. Electron Density can be regarded as the probability of finding an electron in a certain region of space. Since the probability of finding an electron is greater in certain directions than it is in others, we can say that the electron density is greater in these directions. Thus, with p orbitals it can be seen that the electron density is greatest along the x axis for the p_x orbital; similarly, the electron density is greatest along the y axis for the p_y orbital, and along the z axis for the p_z orbital. With s orbitals, the electron density is the same in all directions. This is important when considering the overlap of atomic orbitals.

Orbital Overlap. It has already been established (p. 23) that covalent bonding may be considered as a pairing of electrons brought about by the overlap of two atomic orbitals from different atoms. Each orbital 'contains' one unpaired electron*. We can assume that the strength of a covalent bond is approximately proportional to the degree of overlap, since this can be regarded as increasing the electron density between the two atoms in a bond, which results in a lowering of energy. A large overlap will therefore be associated with the formation of a strong bond. This being so, we can say that bond strength depends to some extent upon the type of orbitals which are overlapped. The s orbitals have, for a given distance, the same electron density in all directions. The p orbitals, on the other hand, must have greatest electron densities along the axes because of their angular probability distributions. Maximum overlap will therefore occur in the direction of these axes. *For a given internuclear distance, p orbitals overlap to a greater extent than do s orbitals of the same principal quantum number.*

The graphical representation of orbitals can now be used to

---

* Strictly speaking, it is incorrect to state that an orbital 'contains' an electron; the electron is 'described by' the orbital.

illustrate the manner of overlap in a few simple molecules which have already been considered in Chapter 3. It is not possible to represent both the angular probability and the radial probability on one diagram. When representing the orbitals diagrammatically we usually consider the angular probability only. This is sufficient because it enables us to see the directions in which the atoms must approach each other to form a bond, and hence we can visualise where there is a build-up of electron density. The diagrams which follow illustrate the ways in which atoms are orientated with respect to one another in order to achieve *the maximum overlap* and therefore *the maximum bond strength*.

Ammonia (see p. 23). The nitrogen atom has the configuration $1s^2 2s^2 2p^3$, or $1s^2 2s^2 2p_x^1 2p_y^1 2p_z^1$.

The angular probabilities of the three p orbitals may be represented as:

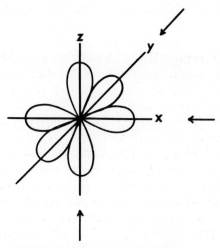

In order for bonding to occur, the three H atoms must approach the N atom along the x, y and z axes (as indicated by the arrows), since in these directions the p electrons are most likely to be found. There will thus be a build-up of electron density between each of the three H atoms and the N atom, and hence three bonds will be formed at right angles to one another, i.e. we should expect the bond angles to be 90°. In fact, the bond angles are 106° 45′.

[Ch. 4]   *The Valence Bond Theory*   47

This deviation may be partly due to repulsive forces between the H atoms.

Water (see p. 24). The oxygen atom has the configuration $1s^2 2s^2 2p^4$, or $1s^2 2s^2 2p_x^2 2p_y^1 2p_z^1$.

The angular probabilities of the three p orbitals are represented below:

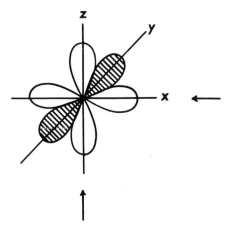

Note. The lone pair orbital is shaded.

This time we need to consider only two of the p orbitals, since the third already contains two electrons. One hydrogen atom will approach the oxygen atom along the x axis to overlap the p_x orbital, and the other will approach along the z axis to overlap p_z.

The two covalent bonds should therefore be formed at right angles to one another. However, the bond angle is in fact 104° 27′ (see p. 34), and this may be partly explained by repulsion from the lone pair and between the hydrogen atoms.

Beryllium Chloride (see p. 25). The configuration of Be is $1s^2 2s^2$, and we have seen how it can be made to produce unpaired electrons by the promotion of one of the 2s electrons into a p orbital to give $1s^2 2s^1 2p_x^1$.

If one bond of BeCl$_2$ is formed by overlapping a p orbital of chlorine with the 2s orbital of beryllium, and the other bond is formed from a p orbital and the beryllium $2p_x$ orbital; then from electron density considerations (see p. 45) we should expect the two bonds to have different strengths. Not only that, but we shall see that the bond angles would also be expected to be indeterminate.

Suppose that we represent the angular probabilities of the $2s2p_x$ orbitals of Be as:

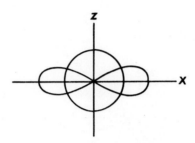

and we consider the manner of overlap by p orbitals of two chlorine atoms (remembering that a Cl atom has an unpaired electron in a $3p$ orbital and supposing that this is the p_x orbital). Then for maximum overlap, one of the Cl atoms and the Be atom will be orientated as shown below:

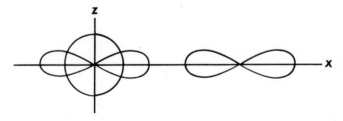

Thus one of the Cl atoms will form a bond with the Be atom by overlap of p_x orbitals, since their angular probabilities are in the same direction.

The other Cl atom, however, could approach from *any direction* since the $2s$ orbital of Be has the same electron density in all directions. The bond angle could therefore have almost any value, for example as shown on page 49.

The p orbital of Be has specific directional properties, whereas the s orbital does not.

The facts are, however, that $BeCl_2$ is a linear molecule, and that the two Be–Cl bond strengths are the same. This brings us

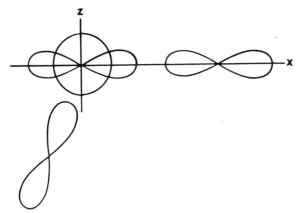

to one of the most important concepts of the valence bond theory, namely, *hybridisation* of atomic orbitals.

Hybridisation. Hitherto we have regarded an electron which has been 'promoted' to another orbital to be described by an orbital of different character from that from which it came. One valency electron of beryllium was taken to be described by an *s* orbital, and the other by a *p* orbital. This is an over-simplification, and according to the valence bond theory the resultant orbitals are a combination of the two. We speak of 'mixing' an *s* and a *p* orbital; but this is entirely in the mathematical sense and cannot be related to any physical process. Mathematical deduction leads us to the conclusion that the *s* and *p* orbitals are *hybridised* to form two equivalent orbitals known as *hybrid orbitals*. These exhibit directional properties, and their angular probabilities are shown diagrammatically below.

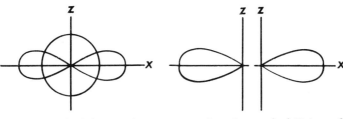

Angular probabilities of *s* and *p* atomic orbitals

Angular probabilities of *sp* hybrid orbitals

We can see that the hybrid orbitals have their maximum probabilities in opposite directions. One electron will tend to be concentrated in the positive direction, and the other in the negative direction, of the x axis.

This type of hybrid is known as an *sp* hybrid orbital; and it is easy to see how molecules such as $BeCl_2$ which are formed from them are *linear*.

Boron has the configuration $1s^2 2s^2 2p^1$, or $1s^2 2s^1 2p_x^1 2p_y^1$ when it forms BF_3 (see p. 27). In this case *three* hybrid orbitals will be formed from *one s* and *two p* orbitals; these are known as *sp^2 hybrid orbitals*, and their angular probability diagrams are shown below:

It can be seen that molecules which are formed from sp^2 hybrids will be *triangular planar*.

Carbon adopts the configuration $1s^2 2s^1 2p_x^1 2p_y^1 2p_z^1$ to form methane (see p. 26), and the *four* hybrid orbitals formed from one *s* and three *p* orbitals are *sp^3 hybrid orbitals*. The maxima for the angular probabilities of these four hybrids are directed towards the corners of a regular tetrahedron. Molecules formed by overlap with sp^3 orbitals are therefore tetrahedral.

Phosphorus $(3s^2 3p^3)$ can show a covalency of five (see p 28), i.e. $3s^1 3p^3 3d^1$; and here we have hybrid orbitals formed from one *s*, three *p* and one *d* orbital, i.e. *five sp^3d hybrid orbitals*. The maxima for the angular probabilities of these five hybrids are directed towards the corners of a trigonal bipyramid. Molecules formed from sp^3d orbitals (e.g. PF_5) are therefore trigonal bipyramidal.

Sulphur $(3s^2 3p^4)$ can show a covalency of six (see p. 30), i.e. $3s^1 3p^3 3d^2$. Here we have hybrid orbitals formed from one *s*, three *p* and two *d* orbitals, i.e. *six sp^3d^2 hybrid orbitals*. The maxima for the angular probabilities of these six hybrids are directed towards the corners of a regular octahedron. Molecules formed from sp^3d^2 hybrid orbitals (e.g. SF_6) are therefore octahedral.

The table below summarises the shapes of molecules formed by the overlap of hybrid orbitals.

Hybrid orbital	Shape of Molecule
sp	Linear
sp^2	Triangular planar
sp^3	Tetrahedral
sp^3d	Trigonal bipyramidal
sp^3d^2	Octahedral

(cf. p. 37)

The concept of hybridisation enables us to explain the bond angles in molecules such as ammonia and water (see pp. 33 and 34). In ammonia, for example, we may consider the valence orbitals of N to be sp^3 hybridised. One of the four sp^3 hybrid orbitals contains a pair of electrons (the lone pair), whilst the other three are singly occupied and form bonds with the three hydrogen atoms, namely:

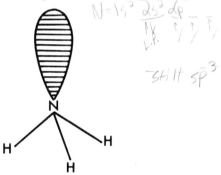

Note. The lone pair sp^3 hybrid orbital is shaded.

The lone pair of electrons is therefore directed towards the fourth corner of the tetrahedron. Since this pair is not localised between two atoms, it may be considered to occupy more space than bonding pairs and hence will repel the other electrons to a greater extent. The bond angle is thus reduced by a small amount.

σ-Bonding

The type of covalent bonding which we have described is characterised by *axial overlap of atomic orbitals*. The combining atoms approach one another along the relevant axes in order to achieve maximum overlap of orbitals, and consequent build-up of electron density. This applies whether or not hybrid orbitals are involved, and results in the formation of strong covalent bonds. Such bonds are known as σ-bonds.

When multiple bonds (as in ethylene for example) are considered, it is found that another type of bond is also involved. We shall now discuss the manner in which these bonds are formed.

CHAPTER 5

π-BONDING

In order to illustrate the mode of combination in molecules which contain multiple bonds, we shall first consider the simple organic compounds: ethane, ethylene and acetylene.

Ethane, C_2H_6, contains only single bonds. The familiar method of writing the formula is:

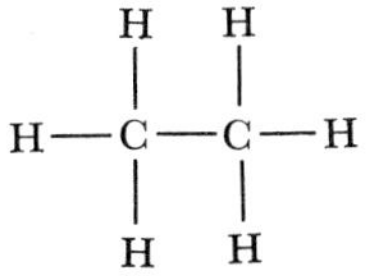

$$\begin{array}{c} HH \\ || \\ H-C-C-H \\ || \\ HH \end{array}$$

The carbon atoms show covalencies of four, and as we have seen in Chapter 4 they are sp^3 hybridised. Each carbon atom uses four sp^3 orbitals; three of these each overlap with an s orbital of hydrogen, and the fourth overlaps the remaining sp^3 of the other carbon atom. Thus, seven σ-bonds are formed; six C–H and one C–C. Now, sp^3 hybridisation leads to the formation of tetrahedral arrangements, so that ethane may be assigned the structure:

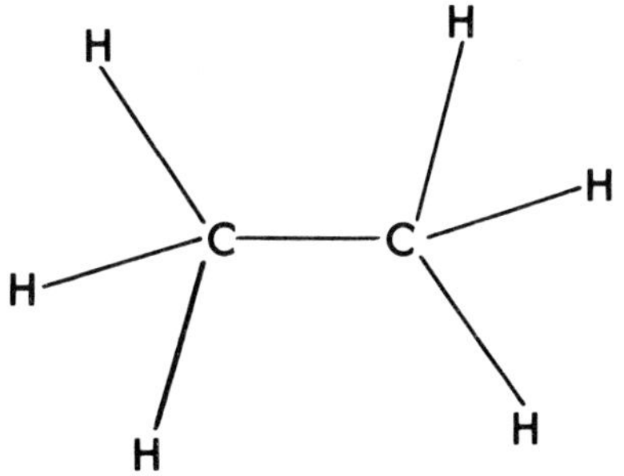

The atoms in molecules, however, are in continual motion. Not only do they exhibit vibrational motion, but they are also able to rotate freely unless restricted in some way. With ethane, rotation about the C–C bond means that the two CH_3 groups are free to rotate relative to one another.

Ethylene, C_2H_4, contains a double C=C bond, and is usually written:

$$\begin{array}{c}H\\ \diagdown\\ C=C\\ \diagup\diagdown\\ HH\end{array}\begin{array}{c}H\\ \diagup\\ \\ \diagdown\\ H\end{array}$$

$C\ 1s^2\ 2s^2\ 2p\ \uparrow\ \uparrow\ \uparrow$

Now, what does this mean in terms of orbital overlap? Each carbon atom is bonded to only three other atoms, and is sp^2 hybridised. Each carbon atom has three sp^2 hybrid orbitals; two of these each overlap with an s orbital of hydrogen, and the third overlaps the remaining sp^2 of the other carbon atom. Five σ-bonds are formed: four C–H and one C–C. Thus sp^2 hybridisation leads to the formation of triangular planar arrangements about each carbon atom.

We know, however, that carbon cannot exhibit a covalency of three (see p. 26). One of the p orbitals of carbon is not used in forming sp^2 hybrid orbitals; therefore two p orbitals (one from each C atom) in ethylene have so far been unaccounted for. These are in the plane which is perpendicular to that of the arrangement:

$$\begin{array}{c}H\\ \diagdown\\ C\text{---}C\\ \diagup\diagdown\\ HH\end{array}\begin{array}{c}H\\ \diagup\\ \\ \diagdown\\ H\end{array}$$

and their angular probabilities are shown below:

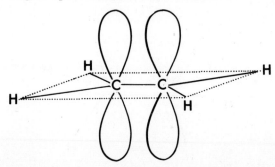

It can be seen that the most effective overlap of p orbitals will not occur in this case. There can be no overlap in the plane of the molecule since there is zero probability of finding either electron along the C–C bond. There will, however, be an increase in electron density both above and below the plane of the molecule; with a region of zero electron density along the bond axis. We could visualise the electron cloud (see p. 5) to be something like this:

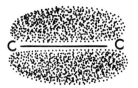

The type of bond which is formed in this manner is called a *π-bond*.

We now see that our C=C double bond comprises one σ-bond and one π-bond. The result of this is to shorten the mean distance between the C nuclei; i.e. the C=C bond length is shorter than that of C–C. Although π-bonds are weaker than σ-bonds, the C=C bonding in ethylene is stronger than the C–C bonding in ethane owing to the combination of σ- and π-bonds.

Rotation of the two carbon atoms with respect to one another is now restricted since such rotation would destroy the π overlap; and this leads to the possibility of *isomerism*. For example, dichloroethylene can have either *cis* or *trans* configurations:

$$\begin{array}{cc} \text{Cl} \diagdown \quad \diagup \text{Cl} & \text{H} \diagdown \quad \diagup \text{Cl} \\ \text{C}=\text{C} & \text{C}=\text{C} \\ \text{H} \diagup \quad \diagdown \text{H} & \text{Cl} \diagup \quad \diagdown \text{H} \end{array}$$

cis-Dichloroethylene   *trans*-Dichloroethylene

Acetylene, C_2H_2, contains a triple C≡C bond: H–C≡C–H.

Here, each carbon atom is bonded to only two atoms, and is sp hybridised. Each carbon has two sp hybrid orbitals; one of these overlaps with an s orbital of hydrogen, and the other overlaps the remaining sp of the other carbon atom. Three σ-bonds are formed; two C–H and one C–C. The molecule is linear.

Each carbon atom has two p orbitals which are not involved in hybridisation; and their angular probabilities are shown on the following page. Two π-bonds will therefore be formed, giving C≡C comprising one σ- and two π-bonds, the bond length being shorter than that of C=C.

The presence of π-bonds has no effect upon the shape of a

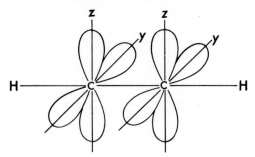

molecule; this is determined solely by σ-bonds and lone pairs of electrons.

We can now consider π-bonding in some simple inorganic molecules.

Nitrogen, N_2, and acetylene are *isoelectronic*; the molecules contain the same total number of orbital electrons, i.e. fourteen.

The angular probabilities of the three p orbitals of the nitrogen atom may be represented as shown on page 46. Suppose that two N atoms are aligned as shown below:

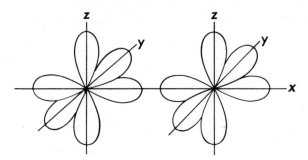

The p_x orbitals of the two atoms will overlap to form a σ-bond, with maximum electron density build-up along the x axis. This leaves the p_y orbitals from each atom to form a π-bond; similarly, the two p_z orbitals will form a second π-bond. Thus, the N≡N molecule has a bonding system consisting of one σ- and two π-bonds. The triple bond of N_2 is very strong, which accounts for the high stability of the molecule.

Carbon dioxide, CO_2, is usually written: O=C=O.

The carbon atom is bonded to only two atoms, and is *sp* hybridised; it is therefore linear. The two *sp* hybrid orbitals form σ-bonds with oxygen; with each *sp* orbital overlapping one

p orbital from each O atom. This leaves two p orbitals of carbon and one p orbital from each oxygen atom with unpaired electrons, and thus two π-bonds will be formed, one between carbon and each oxygen atom. Each double bond of CO_2 consists of one σ- and one π-bond, a total of two σ- and two π-bonds.

A discrepancy is found to occur, however, in the molecule of carbon dioxide. The C=O bond length is shorter than would be expected and shows characteristics intermediate between those of a double and a triple bond. This is explained by the concept of *resonance*.

Resonance. So far, we have considered a covalent bond to be formed between two atoms by the equal sharing of electron pairs, but this is not necessarily the case (see p. 38). Also, we have assumed that bonds are either single, double or triple. This simple theory does not always give an accurate picture of the bonding in certain molecules, and can be considered to be a weakness in the valence bond theory. CO_2, for example, would be expected to have the structure O=C=O; but as we have remarked, the bonds are apparently intermediate in character between a double and a triple bond. The normal C=O bond length is 1·22 Å; and the C≡O bond length is 1·10 Å. In CO_2, however, the carbon-oxygen bond is found to have a length of 1·15 Å. Clearly, the real structure must be somewhere between the two.

The theory of resonance postulates that bonds may be intermediate in character between certain hypothetical structures (often called *canonical forms*). For example, the molecule of carbon dioxide can be considered to have a structure intermediate between the three hypothetical forms:

(i)   O=C=O
(ii) $^+$O≡C–O$^-$
(iii) $^-$O–C≡O$^+$

In fact, no molecule of CO_2 has any of these structures; they describe only the extreme forms which can be pictured non-mathematically.

Carbon dioxide is said to be a *resonance hybrid* of the above canonical forms. The resonance hybrid has a *lower energy* than any of the canonical forms; and the difference in energy between it and the most stable of the hypothetical structures is called the *resonance energy*. The relevant energy levels for CO_2 are shown diagrammatically below:

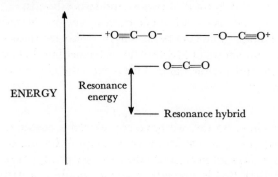

Structures (ii) and (iii) require a little explanation. Ignoring the σ-bonds, and regarding only the p orbitals which are available for π-bonding, we have angular probabilities as follows:

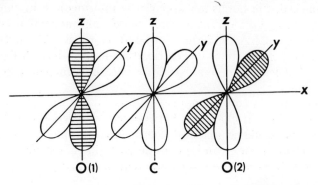

Note. The lone pair angular probabilities are shaded.

Now π-bonds can be formed between the p_y orbital of $O_{(1)}$ and the p_y orbital of C, and between the p_z orbital of C and the p_z orbital of $O_{(2)}$. As the electronic configurations stand, however, a triple bond cannot occur between C and O, since the remaining p orbitals of both oxygen atoms already have paired electrons (i.e. lone pairs). In order to make it possible, one electron from the p_z orbital of $O_{(1)}$ must be transferred to the p_z orbital of

[Ch. 5] π-Bonding 59

$O_{(2)}$. Two π-bonds can then be formed between C and $O_{(1)}$ leaving only a σ-bond between C and $O_{(2)}$. Thus we have:

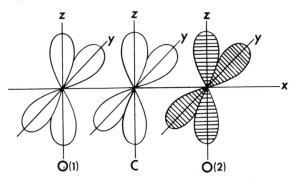

$O_{(1)}$ has therefore *lost* an electron and becomes *positive*; and $O_{(2)}$ has *gained* an electron to become *negative*. This gives structure (ii).

Structure (iii) would be given if one electron from the lone pair p_y orbital of $O_{(2)}$ were lost to the p_y orbital of $O_{(1)}$.

In order for the hypothetical forms to contribute significantly to a resonance hybrid, they must have certain properties in common:

(i) They must all have the same number of unpaired electrons.
(ii) The constituent atomic nuclei must occupy the same positions relative to one another.
(iii) They must all have similar energies.

Many other molecules and ions exhibit resonance, as in the following examples.

The nitrite ion, NO_2^-, has hypothetical forms:

The carbonate ion, CO_3^{2-}, is a resonance hybrid of:

Introduction to Valence Theory

Dinitrogen tetroxide, N_2O_4, is a resonance hybrid of:

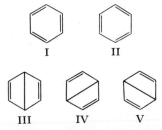

Benzene, C_6H_6, is a classic example of how an application of the theory of resonance enables us to understand otherwise inexplicable properties of the molecule. These include:

(i) Extreme stability.
(ii) A completely planar structure.
(iii) The same bond length between carbon atoms (i.e. the molecule is a regular hexagon of bond angle 120°).
(iv) Characteristic chemical reactions of substitution (e.g. the formation of C_6H_5Br) rather than additions as is the case with most unsaturated (double-bonded) aliphatic compounds (e.g. the formation of ethylene dibromide, $C_2H_4Br_2$).

There are five canonical forms which contribute to the resonance hybrid:

Structures I and II were proposed by Kekulé and contribute to a greater extent than do structures III, IV and V, which were proposed by Dewar (these would be less stable owing to the presence of weak bonding across the benzene ring).

Each carbon atom is bonded to three atoms, and is sp^2 hybridised. The three sp^2 orbitals of each C atom form σ-bonds with two other carbon and one hydrogen atom, to form the system:

```
        H
        |
   H    C    H
    \  / \  /
     C    C
     |    |
     C    C
    /  \ /  \
   H    C    H
        |
        H
```

which can be represented as:

This leaves each C atom with one *p* orbital which is not involved in σ-bonding, and these are at right angles to the plane of the molecule. Their angular probability representations are given below:

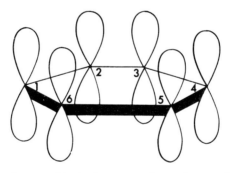

Suppose that the *p* orbital of $C_{(1)}$ overlaps the *p* orbital of $C_{(2)}$; that of $C_{(3)}$ overlaps that of $C_{(4)}$; and that of $C_{(5)}$ overlaps that of $C_{(6)}$. This can be said to represent structure I.

Similarly, structure II would be formed by the overlap of orbitals from $C_{(2)}$ and $C_{(3)}$; $C_{(4)}$ and $C_{(5)}$; and $C_{(6)}$ and $C_{(1)}$.

Structures III, IV and V can be worked out in the same way.

Mathematical solution of these five hypothetical forms would then give a general picture of the resonance hybrid.

There is another way of looking at this, however. Since the distance between each carbon atom is the same, it is impossible for the *p* orbital of one C atom to overlap that of only *one* of its neighbours without also overlapping that of its *other* neighbour. Overlap of orbitals one with another will therefore occur around the benzene ring. In other words, two 'rings' of electron density build-up will be formed, one *above* and one *below* the plane of the molecule:

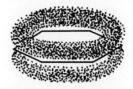

This is known as a *delocalised π-electron system*, and leads to the formation of bonds which are intermediate in character between single and double bonds. It accounts for the fact that benzene does not exhibit the properties of molecules which possess normal double bonds (such as ethylene). Also, if a system of single and double bonds existed (such as in any one of the canonical forms), the carbon–carbon bond lengths would not all be equal. The new orbital system so formed encompasses the whole molecule. In other examples for which we have used the valence bond theory, we have been concerned only with the overlap of orbitals from adjacent atoms. With benzene, however, we have taken a step forward and have postulated an orbital which is characteristic of the molecule as a whole, i.e. a *molecular orbital*. In this alternative way of looking at the bonding system in benzene, *there has been no need to take into account the various canonical forms of the resonance hybrid.*

The concept of resonance is very useful in the light of the valence bond theory; but for more complex molecules (particularly for large aromatic systems), the number of canonical forms (and consequent mathematical solution of each one) becomes unmanageable. The theory of molecular orbitals proves to be more satisfactory. It can also be used successfully for all other molecules, and will be given more consideration in the next chapter.

CHAPTER 6

MOLECULAR ORBITALS

The valence bond theory is concerned with atomic orbitals which do not lose their identity when involved in chemical bonding. The *molecular orbital theory* postulates the formation of new orbitals when atoms combine with one another; the individual atomic orbitals are replaced by orbitals which are characteristic of the molecule as a whole.

Some of the properties of molecular orbitals are:
1. Each orbital is associated with *all* the atomic nuclei of which the molecule is composed.
2. A molecular orbital is described by mathematical combination of the individual atomic orbitals concerned.
3. A bonding molecular orbital has a *lower energy* than have the individual atomic orbitals from which it is formed. The electron associated with it is therefore more stable than in an isolated atom.
4. As with atomic orbitals (see p. 9), each molecular orbital has a certain energy level. Electrons are 'fed in' to the available molecular orbitals in the order of increasing energy, i.e. the orbital of lowest energy is filled first. A maximum of two electrons can occupy one molecular orbital.

We shall now consider the bonding in diatomic molecules from the viewpoint of the molecular orbital theory.

Bonding in Diatomic Molecules. It is important to stress that for bonding to occur there must be an overall lowering of energy when atoms combine to form a molecule.

Detailed mathematics are involved in calculating the energy levels of molecular orbitals. But we can arrive at a set of basic rules which will enable us to use the theory without involving a description of such calculations.

1. The conservation of orbitals: Combination of two atomic orbitals gives two molecular orbitals.

2. One of these molecular orbitals has a *lower energy* than have the individual atomic orbitals from which it is derived. The other molecular orbital has a *higher energy* than have the atomic orbitals.
3. The energy difference between *each* molecular orbital and the atomic orbitals is approximately the same, i.e.

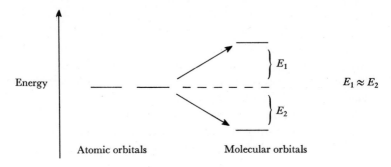

4. Since bonding is associated with an overall lowering of energy, the molecular orbital which has the lower energy will be used in the formation of a chemical bond. This is called the *bonding molecular orbital*. The other molecular orbital has a higher energy and will not be used to form a bond (since it would be less stable), and is called an *antibonding molecular orbital*.

These rules are summarised in the diagram below:

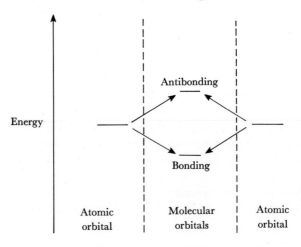

A bonding molecular orbital is associated with a build-up of electron density (i.e. lowering of energy); whereas an antibonding molecular orbital is associated with a decrease of electron density (i.e. increasing energy).

As in the valence bond theory, both σ- and π-type bonding can occur.

We are now in a position to consider the bonding of elements from the first two periods of the periodic table.

Hydrogen

$$2H \longrightarrow H_2$$

The electronic configuration of H is $1s^1$. There is a total of two electrons to be fed into molecular orbitals. Both of these will occupy the bonding molecular orbital as shown below. This is known as the σ1s molecular orbital, and the configuration of H_2 can be written $(\sigma 1s)^2$; i.e. two electrons in the σ1s orbital.

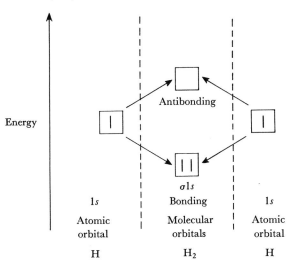

Note. Each 'box' represents an orbital, and the vertical lines therein represent electrons.

Helium $(1s^2)$. Let us now consider whether there is the possibility of forming a molecule He_2.

The same molecular orbitals are available as for hydrogen, but we now have four electrons to feed into them. Two will occupy the bonding molecular orbital, and the other two will occupy the antibonding molecular orbital (designated σ^*1s).

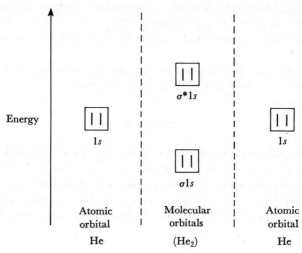

There will be no net lowering of the energy of the system, and hence no stabilisation. A molecule will not be formed therefore.

However, in an electrical discharge tube, the He_2^+ ion can be formed. In this case there are only three electrons, two of which occupy the bonding molecular orbital, and one the antibonding molecular orbital. There will therefore be a small lowering of the overall energy, and hence some stabilisation. He_2^+ has the configuration $(\sigma 1s)^2(\sigma^*1s)^1$.

Lithium $(1s^22s^1)$

$$2Li \longrightarrow Li_2$$

The $1s$ electrons are considered to remain in the atomic orbitals, since they are more firmly held than are electrons in the valency shell. Two molecular orbitals will thus be formed from the two $2s$ atomic orbitals (one from each Li atom).

The two electrons will occupy the $\sigma 2s$ bonding molecular orbital (see p. 67).

Li_2 has the configuration $(\sigma 2s)^2$. It is found to exist in the vapour state.

Beryllium $(1s^22s^2)$. Is there a possibility of forming Be_2? As with Li, the $1s$ electrons are taken as remaining in the atomic orbitals. Four valency electrons must be considered (two from

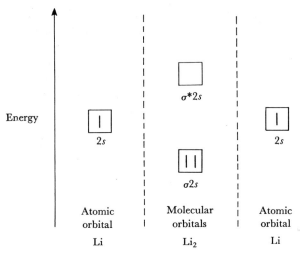

each Be atom). Two of these will occupy the σ2s bonding molecular orbital, and the other two will occupy the σ*2s antibonding molecular orbital.

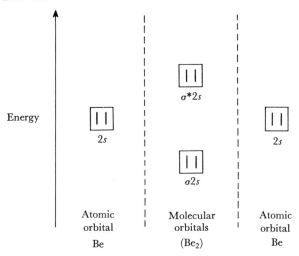

There will therefore be no net lowering of the energy, and so a molecule will not be formed.

For the next elements in Period 2 we must take into consideration molecular orbitals formed from p as well as s atomic orbitals. We have seen on page 56 that if the p_x orbitals of combining atoms overlap to form a σ-bond then the p_y orbitals of the atoms will form a π-bond, and the p_z orbitals will also form a π-bond. Similarly, p_x atomic orbitals of each atom will form σ-type molecular orbitals, p_y orbitals from each atom will form π-type molecular orbitals, and the p_z orbitals will also form π-type molecular orbitals. From the conservation of orbitals there will be four π-type molecular orbitals formed—two bonding (designated π_y and π_z) and two antibonding (π_y^* and π_z^*).

The relative energies of the molecular orbitals which are formed from 2s and 2p atomic orbitals are shown below:

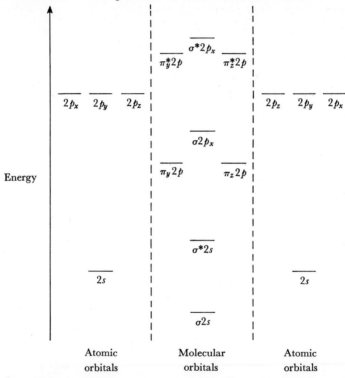

You will notice that the π_y and π_z (also the π_y^* and π_z^*) molecular orbitals have equal energies. This is because the p atomic orbitals from which they are derived have equal energies; the only difference between them is their orientation in space.

Now we shall discuss the bonding between atoms which contain 2p electrons.

Boron ($1s^2 2s^2 2p^1$)

$$2B \longrightarrow B_2$$

There are six valency electrons (three from each B atom) to be accommodated in molecular orbitals. Two of these will occupy the $\sigma 2s$ bonding molecular orbital, and two will occupy the $\sigma^* 2s$ antibonding molecular orbital. The remaining two electrons will go into the molecular orbitals of next lowest energy—i.e. π-type molecular orbitals. We have seen (p. 11) that when filling atomic orbitals of equal energy, the most stable configuration is when the electrons are in different orbitals (i.e. when they are unpaired). This is also true for electrons in molecular orbitals. With B_2, therefore, one electron will go into the π_y molecular orbital and one into the π_z molecular orbital.

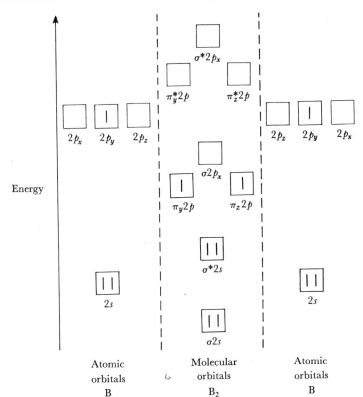

The configuration of B_2 is thus $(\sigma 2s)^2(\sigma^*2s)^2(\pi_y 2p)^1(\pi_z 2p)^1$.
Carbon $(1s^2 2s^2 2p^2)$

$$2C \longrightarrow C_2$$

There are eight valency electrons (four from each C atom) to be accommodated in molecular orbitals. Two of these occupy the $\sigma 2s$ molecular orbital and two occupy the σ^*2s molecular orbital. The four remaining electrons are accommodated in the $\pi_y 2p$ and $\pi_z 2p$ molecular orbitals—two in each. The configuration of C_2 is therefore $(\sigma 2s)^2(\sigma^*2s)^2(\pi_y 2p)^2(\pi_z 2p)^2$.

Three pairs of electrons are in bonding molecular orbitals and one pair is in an antibonding molecular orbital. The net number of electron pairs involved in bonding is *two*; this means that there is a double bond in the C_2 molecule, i.e. C=C.

Nitrogen $(1s^2 2s^2 2p^3)$

$$2N \longrightarrow N_2$$

The ten valency electrons occupy the molecular orbitals as shown on page 71. The configuration of N_2 is therefore:

$$(\sigma 2s)^2(\sigma^*2s)^2(\pi_y 2p)^2(\pi_z 2p)^2(\sigma 2p_x)^2$$

Four pairs of electrons are in bonding molecular orbitals and one pair is in an antibonding molecular orbital; i.e. the net number of electron pairs involved in bonding is *three*, giving a triple bond.

Obviously the N≡N bond is extremely strong. Of the ten valency electrons, eight are in bonding molecular orbitals. Also, all the $2p$ bonding molecular orbitals and none of the antibonding molecular orbitals are filled.

Oxygen $(1s^2 2s^2 2p^4)$.

$$2O \longrightarrow O_2$$

The oxygen molecule has two more electrons than the nitrogen molecule. One of these will occupy the π_y^*2p molecular orbital and one will occupy the π_z^*2p molecular orbital. There are therefore two unpaired electrons. The presence of unpaired electrons gives rise to the property of *paramagnetism* (see p. 80) which is exhibited by O_2, and which can be adequately explained only by the molecular orbital theory.

[Ch. 6] *Molecular Orbitals* 71

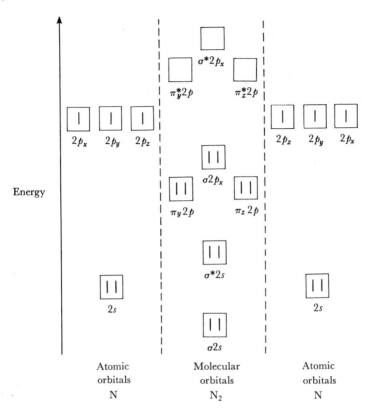

There are eight electrons in bonding molecular orbitals and only four in antibonding molecular orbitals. The O=O bond is therefore strong.

It is interesting to note that if an electron is removed to form O_2^+, this will come from an antibonding π^* molecular orbital (since electrons in these orbitals have the highest energies) and will result in even greater stability. There is a compound $O_2^+PtF_6^-$ which was the forerunner of the first known noble gas compound. It was noticed that O_2 and Xe have the same ionisation potential (see pp. 17–18) and this suggested that PtF_6 (a powerful oxidising agent) could possibly oxidise Xe. The compound $XePtF_6$ was thus isolated!

The configuration of O_2 is

$$(\sigma 2s)^2(\sigma^*2s)^2(\sigma 2p_x)^2(\pi_y 2p)^2(\pi_z 2p)^2(\pi_y^*2p)^1(\pi_z^*2p)^1$$

Thus the $2p_x$ molecular orbital is filled before the π-type molecular orbitals, since for O_2 and F_2 it has a lower energy.

Fluorine ($1s^2 2s^2 2p^5$)

$$2F \longrightarrow F_2$$

F_2 has two more electrons than O_2. One of these will go into the $\pi_y^* 2p$ molecular orbital and the other will go into the $\pi_z^* 2p$ molecular orbital. The configuration of F_2 is therefore:

$$(\sigma 2s)^2 (\sigma^* 2s)^2 (\sigma 2p_x)^2 (\pi_y 2p)^2 (\pi_z 2p)^2 (\pi_y^* 2p)^2 (\pi_z^* 2p)^2$$

Neon ($1s^2 2s^2 2p^6$). If a molecule Ne_2 were formed, then all the molecular orbitals would be filled. The number of electrons in bonding molecular orbitals would be equal to the number in antibonding molecular orbitals. No compound is formed therefore.

Bond Order and Bond Strength. From this application of the molecular orbital theory we can deduce the bond multiplicity (or *bond order*), and also the relative bond strengths of the diatomic molecules considered.

Bond order is defined as:

$\frac{1}{2}$ (no. of electrons in bonding molecular orbitals −

no. of electrons in antibonding molecular orbitals).

The bond order of H_2 $(\sigma 1s)^2$ is therefore *one* (i.e. a single bond, H–H). This is strong, since there are no electrons in antibonding molecular orbitals; also, there are no inner shell electrons to set up additional repulsive forces.

The bond order of He_2^+ $(\sigma 1s)^2 (\sigma^* 1s)^1$ is 0·5 and the bond is extremely weak.

Li_2 $(\sigma 2s)^2$ has a bond order of one. The molecule is less stable than H_2 owing to repulsion from the four inner shell electrons.

B_2 $(\sigma 2s)^2 (\sigma^* 2s)^2 (\pi_y 2p)^1 (\pi_z 2p)^1$ has a bond order of one and the bond strength is between that of Li_2 and H_2.

C_2 $(\sigma 2s)^2 (\sigma^* 2s)^2 (\pi_y 2p)^2 (\pi_z 2p)^2$. As we have seen, the bond order is two. The C=C bond is considerably stronger than is a single bond.

N_2 $(\sigma 2s)^2 (\sigma^* 2s)^2 (\pi_y 2p)^2 (\pi_z 2p)^2 (\sigma 2p_x)^2$ has already been discussed. The triple bond is extremely strong, and N_2 has the highest bond strength of all the diatomic elements of the first and second periods.

O_2 $(\sigma 2s)^2(\sigma^*2s)^2(\sigma 2p_x)^2(\pi_y 2p)^2(\pi_z 2p)^2(\pi_y^* 2p)^1(\pi_z^* 2p)^1$ has a bond order of two. The bond strength is considerable, but is less than that of C=C. which has no electrons in antibonding $2p$ molecular orbitals.

F_2 $(\sigma 2s)^2(\sigma^*2s)^2(\sigma 2p_x)^2(\pi_y 2p)^2(\pi_z 2p)^2(\pi_y^* 2p)^2(\pi_z^* 2p)^2$ has a bond order of one. The bond strength is less than would be expected for a single bond. This is due to repulsion from inner shell electrons. Also, E_1 (see p. 64) for the π^* molecular orbitals is somewhat higher than E_2 for the π molecular orbitals. Contribution from the antibonding molecular orbitals will therefore be greater than if $E_1 = E_2$.

Bonding between atoms of different elements becomes rather more complex, and a detailed treatment by means of the molecular orbital theory is beyond the scope of this book.

One application of the theory must be mentioned however. There are certain compounds for which no satisfactory classical structure can be drawn, and for which no adequate explanation can be made by means of simple electron pair bonds. The hydrides of boron are such compounds, and we shall illustrate the type of bond which is formed by considering diborane.

The Three-Centre Bond. The molecule of diborane, B_2H_6, has two boron atoms and four hydrogen atoms in the same plane. The other two hydrogen atoms are located one above and one below this plane, and in between the two boron atoms. The hydrogen atoms are placed in approximately tetrahedral positions about each boron atom as shown:

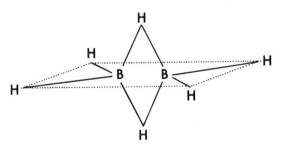

The structure of the molecule implies that each B atom forms four bonds with H atoms.

The electronic configuration of B is $1s^2 2s^2 2p^1$: there are thus *three* valency electrons available from each atom. This is insufficient for B to form electron pair bonds with *four* hydrogen atoms, and the molecule is said to be *electron deficient*. In BF_3 for example, B is sp^2 hybridised (see p. 50), giving only three atomic orbitals instead of (for diborane) the required four. Molecular orbitals, however, can be spread over more than two atomic nuclei, and there is no requirement that two electrons must be shared between two adjacent atoms.

We have seen that each boron atom has four valence orbitals—one s and three p; and the geometry of the molecule suggests that these orbitals are sp^3 hybridised. The angular probability diagrams of the hybrids are shown below, together with those of the 1s orbital from each hydrogen atom which are positioned so as to obtain maximum overlap. The electron associated with each orbital is schematically shown as 'e'.

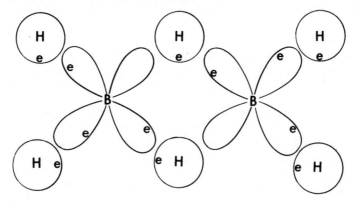

The terminal B–H bonds present no problem. These four σ-bonds are each formed by the overlap of a hybrid orbital containing one electron with a hydrogen 1s orbital, which also contains a single electron. They are 'normal' covalent bonds, which are called *two-centre two electron bonds* because the bond is formed between only two atoms.

The two 'bridge' bonds are different, however. In this case each bond is formed by the overlap of *two* boron sp^3 hybrid orbitals with a 1s orbital of hydrogen, i.e. a molecular orbital is formed which extends over the *three* atoms. It can be seen that there are just the required two electrons to fill this molecular orbital. Since the orbital extends over three atoms and contains two electrons, this type of bond is known as a *three-centre two electron bond*. The electron clouds associated with these bonds are shown schematically below, and illustrate why they are sometimes called 'banana bonds'.

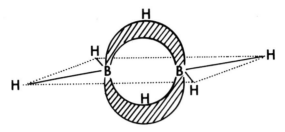

We have now covered the major types of bonding which occur between atoms when they form molecules. There are other, less well defined, chemical bonds; but before proceeding to discuss these we shall apply the principles which have been developed to one or two interesting compounds.

CHAPTER 7

BONDING IN NOBLE GAS COMPOUNDS

It has already been mentioned that until recently the elements of Group **0** were thought to be completely inert, and incapable of entering into chemical combination with other elements. More than thirty years ago, however, it was predicted from the valence bond theory that it should be possible to form compounds between the noble gases and, for example, fluorine; but it is only during the past few years that such compounds have been isolated.

We have seen how the first noble gas compound, $XePtF_6$, came to be prepared (p. 71); and there are now many more, of which we shall consider only the fluorides of Xe, i.e. XeF_2, XeF_4 and XeF_6.

The shapes of these molecules can be predicted from the simple electron pair repulsion theory (p. 37).

XeF_2. The total number of electrons in the valence shell on compound formation is ten, i.e. 8 (from Xe) + 2 (one from each F). These five pairs of electrons would be expected to be directed towards the corners of a trigonal bipyramid (see p. 34), with the fluorine atoms occupying the axial positions (see p. 35):

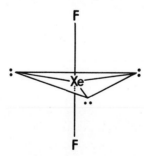

The molecule will therefore be linear.

XeF_4. There are twelve valence electrons, 8 (from Xe) + 4 (one from each F). The electron pairs are directed towards the

when the substance is *ferromagnetic*; or in antiparallel:

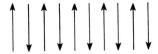

when the substance is *antiferromagnetic*. Usually, however, a random orientation is found.

Some compounds in which the metal ion is in the same valence state have very different magnetic moments. For example, $[Co(NH_3)_6]^{3+}$ is diamagnetic, whilst CoF_6^{3-} is strongly paramagnetic.

A satisfactory bonding theory must be able to explain these facts.

Application of the Valence Bond Theory. This approach to the bonding in transition metal compounds was developed by Linus Pauling. We shall consider only elements from the first series of transition metals: scandium, titanium, vanadium, chromium, manganese, iron, cobalt, nickel and copper. The configurations of the valence electrons are:

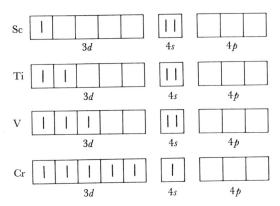

Note that this configuration is more stable than

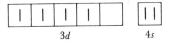

As a general rule, half-filled and completely filled shells are stable states.

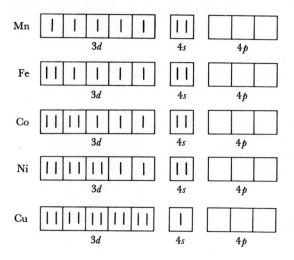

In forming ions, the 4s electrons are lost first. For example:

Fe^{2+} has six d electrons, i.e. d^6
Fe^{3+} has the configuration d^5
Co^{3+} has the configuration d^6
and Cu^{2+} has the configuration d^9

Examination of the electronic configurations reveals that many of the $3d$, $4s$ and $4p$ valence orbitals are unoccupied. It would seem reasonable to suppose, therefore, that a large part of transition metal chemistry involves compounds which are formed by the metal ions accepting electrons donated by electron donors. We have already seen how this occurs with boron trifluoride when the vacant orbital of boron accepts electrons from ammonia to give the coordination compound $H_3N \rightarrow BF_3$ (see p. 27). The electron pair donors are known as *ligands*, which may be either neutral molecules such as ammonia, or ions such as fluoride, chloride, cyanide etc. The resulting compounds are called *coordination complexes*.

[Ch. 8] *Bonding in Transition Metal Compounds* 83

The bonds which are formed by overlapping orbitals of a metal ion (or atom) and a ligand are assumed to be covalent; and the vacant orbitals on the metal which accept the electrons must be hybridised to form the correct shape of the molecule or complex ion. The number of ligands around a particular metal atom or ion is called the *coordination number*. The complexes may be either positively or negatively charged, or they may be neutral, depending upon the metal and the ligand involved. For example, in $[Co(NH_3)_6]^{3+}$ $3Cl^-$ the complex ion consists of the ion Co^{3+} and six neutral ammonia molecules; the overall charge being 3+. The complex ion in $K_3[CoF_6]^{3-}$ consists of Co^{3+} and $6F^-$, giving an overall charge of 3−. In $Fe(CO)_5$, Fe and 5CO are neutral.

The most common coordination number is six, and the most common shape of the complexes is octahedral (i.e. the ligands are arranged at the corners of an octahedron, with the metal atom or ion in the centre).

An example with a coordination number of five is $Fe(CO)_5$, which has a trigonal bipyramidal shape. Coordination complexes with a coordination number of four may be either tetrahedral (e.g. $CoCl_4^{2-}$), or square planar (e.g. $[Cu(NH_3)_4]^{2+}$).

We shall now consider a few examples of octahedral coordination complexes to illustrate the way in which bonding can occur.

$[Ti(H_2O)_6]^{3+}$. Ti^{3+} has the configuration d^1, i.e.

The six water molecules each donate a pair of electrons. In order to obtain an octahedral arrangement, the vacant metal ion orbitals must form hybrids d^2sp^3. Now d^2sp^3 hybridisation involves 'inner' d orbitals (in this case $3d$) and outer shell s and p orbitals ($4s$ and $4p$); the shape of molecules formed by overlap with these hybrids is the same as with sp^3d^2 hybrid orbitals (see p. 51). We may thus represent the formation of $[Ti(H_2O)_6]^{3+}$ schematically as follows:

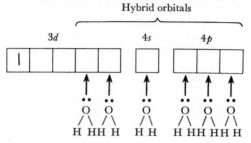

There is one unpaired electron present.

[Cr(NH$_3$)$_6$]$^{3+}$. With ammonia, Cr^{3+} forms the complex [Cr(NH$_3$)$_6$]$^{3+}$. The magnetic moment corresponds to the presence of three unpaired electrons.

Cr^{3+} has the configuration d^3; and in the complex, d^2sp^3 hybrid orbitals will be involved as indicated below.

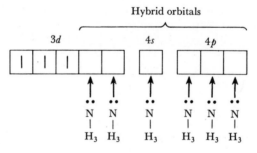

Now some metals form complexes which differ in magnetic moment although in each compound the metal is in the same valence state. For example, magnetic measurements show that the complex ion [CoF$_6$]$^{3-}$ has four unpaired electrons, whilst [Co(NH$_3$)$_6$]$^{3+}$ has no unpaired electrons.

Co^{3+} has the configuration d^6, i.e.

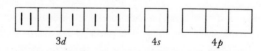

In order to form d^2sp^3 hybrid orbitals the electrons in the $3d$ orbitals must all be paired up, and this will leave two vacant d

orbitals which may then form the required hybrids to accept electrons:

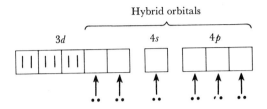

There will thus be no unpaired electrons.

For the complex with four unpaired electrons, $(CoF_6)^{3-}$, the $3d$, orbitals are already occupied; and in order to form an octahedral complex, outer shell $4d$ orbitals must be used to form the hybrids sp^3d^2. Formation of the complex is shown schematically below:

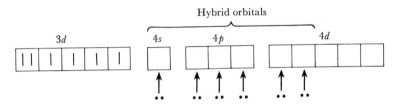

Complexes in which $3d$ orbitals are involved are known as *inner orbital complexes*; whilst the term *outer orbital complex* is used for complexes in which $4d$ orbitals are involved. Outer orbital complexes tend to be less stable than inner orbital complexes because of the high energy of the $4d$ orbitals. The high energy which would be involved in the formation of such complexes throws some doubt upon the validity of the valence bond theory as applied to transition metal compounds. In addition, it is not easily seen how the colour variations of the complexes arise. A more satisfactory explanation of the bonding is given by the *crystal field theory*.

Electrostatic Crystal Field Theory. This theory is very different from the valence bond theory, which starts off with the premise that the bonding is covalent. The crystal field theory in its simplest form completely neglects covalent bonding, and assumes that the bond between metal ion and ligand is totally ionic. The ligands are considered as *point negative charges*, being either negative ions such as fluoride, or *ion dipoles* (for example the lone pair of electrons in ammonia).

To understand the electrostatic theory, we must have a clear idea of the spatial orientation of the d orbitals, which have the following angular probabilities:

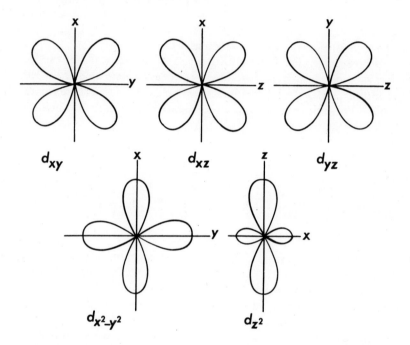

The significance of these diagrams is that for three of the orbitals there is zero probability of finding the electron along the axes, and maximum probability of finding it between the axes. These are called the d_ϵ subset. For the two other orbitals there is a maximum probability of finding the electron along the axes, and these are called the d_γ subset. For the d_ϵ subset then, the electron density will be greatest between the axes; and for the d_γ subset it will be greatest along the axes.

We shall now consider the formation of an octahedral complex, assuming that the ligands are negative charges approaching the metal ion along the axes. Thus we have:

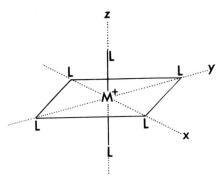

Since the ligands lie along the axes and are negatively charged, any electrons in the d_γ orbitals will experience far more electronic repulsion than will electrons in the d_ϵ orbitals, which lie between the axes.

In the free metal atom or ion the five d orbitals are in a *spherical environment* (i.e. they are free from external electrostatic influences). Also, they have the same energy (orbitals with equal energies are said to be *degenerate*). When the metal atom or ion becomes involved in complex formation as above (i.e. it is in an octahedral environment or *octahedral field*), the orbitals are raised in energy; the d_γ more than the d_ϵ orbitals. The original five degenerate d orbitals split into the two subsets: d_ϵ, which is *triply degenerate* (i.e. there are three orbitals of equal energy); and d_γ, which is *doubly degenerate* (two orbitals of equal energy). The energy difference between the two subsets is given by Δ (p. 88).

Formation of the complex results in a *decrease in symmetry*; the spherical environment associated with the free atom or ion has now been replaced by an octahedral environment. *A decrease in symmetry results in a decrease in degeneracy.*

The magnitude of Δ depends upon both the positive charge on the metal ion and the nature of the ligand.

The bonding then, is electrostatic.

We shall now consider how the electrons occupy the d orbitals in complexes, where they are not completely degenerate. To illustrate this, we shall take the aquo complexes $[Ti(H_2O)_6]^{3+}$, $[V(H_2O)_6]^{3+}$ and $[Cr(H_2O)_6]^{3+}$ which involve one, two and three d electrons respectively.

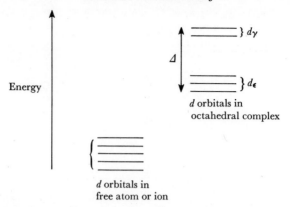

d orbitals in octahedral complex

d orbitals in free atom or ion

The electrons will occupy the subset of lowest energy first, i.e. the d_ϵ orbitals, and will enter the orbitals separately:

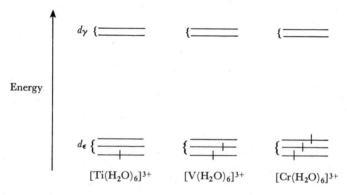

The magnetic moments will therefore correspond to one, two and three unpaired electrons respectively.

For metal ions with more than three electrons, there are two possibilities. For a d^4 system the fourth electron may enter either the d_γ subset or the d_ϵ subset (as shown on p. 89). Which one of these configurations is adopted depends upon the magnitude of Δ. If Δ is greater than the energy required to pair an electron in a d orbital, the configuration will be II; if not, then it will be I. If situation I results, the field is known as a weak crystal field; and the complex is called a *spin free complex*—i.e. the number of unpaired electrons in the complex is the same as in the free ion. This is equivalent to an outer orbital complex as postulated by the valence bond theory. If II occurs, the field is

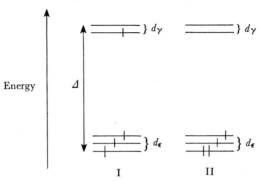

said to be strong and the complex is known as a *spin paired complex*—i.e. the number of unpaired electrons is less than in the free ion. This is equivalent to an inner orbital complex.

As stated previously, the strength of the crystal field depends upon both the metal ion and the ligand; but for a particular metal ion, the ligands may be arranged in order of increasing strength:

$$Cl^- < F^- < H_2O < NCS^- < NH_3 < NO_2^- < CN^- < CO$$
weak $\qquad\qquad\qquad\qquad\qquad\qquad\qquad$ strong

For example, Fe^{3+} which has five d electrons can form complexes: $[FeF_6]^{3-}$ and $[Fe(CN)_6]^{3-}$. The former will be associated with a weak crystal field, and the latter with a strong crystal field:

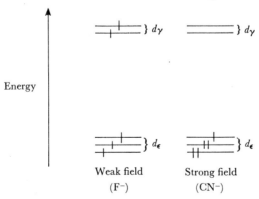

These two complexes, which both contain Fe^{3+}, have very different magnetic moments; $[FeF_6]^{3-}$ is spin free, and $[Fe(CN)_6]^{3-}$ is spin paired.

Another example is Co^{3+}, which has six d electrons, and forms complexes: $[Co(NH_3)_6]^{3+}$ and $[CoF_6]^{3-}$. The former is diamagnetic, while the latter has a magnetic moment corresponding to four unpaired electrons:

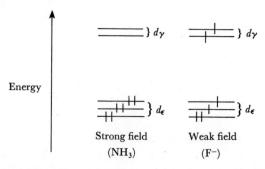

Thus, from this simple theory it is easy to see how spin free and spin paired complexes occur.

Colours of Transition Metal Compounds. The crystal field theory is of great use in understanding the colours of many of the transition metal compounds.

Since the Δ values are relatively small, it is possible to promote (or *excite*) electrons from the lower d_ϵ subset of orbitals to the upper d_γ subset by ordinary visible light. The condition that electrons may be excited is that the difference in energy between the two subsets, Δ, must be equal to $h\nu$; where ν is the frequency of radiation, and h is Planck's constant.

For the complex $[Ti(H_2O)_6]^{3+}$, Δ is about 57 kcal (248 kJ), and requires radiation of wavelength approximately 5000 Å (i.e. in the visible region of the spectrum) in order to excite the electron from d_ϵ to d_γ:

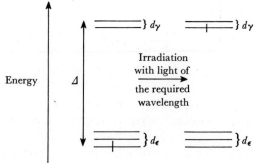

If the light absorbed during this process is plotted against the wavelength, a *spectrum* for $[Ti(H_2O)_6]^{3+}$ is obtained:

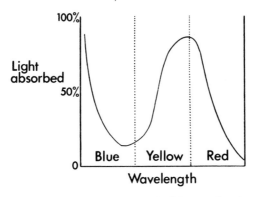

Mainly yellow light is absorbed and the blue and red transmitted; the compound will therefore be purple.

Note that the absorption peak is very broad. It might have been expected to be sharp, since we have considered Δ to be constant. The reason is that the ligands are in continuous vibration, so that their relative positions with respect to the metal ion change slightly; thus they vary with time, which causes the light to be absorbed over a range of wavelengths.

Tetrahedral Complexes. The diagram below shows the tetrahedral arrangement of ligands about a metal ion with respect to the axes.

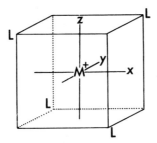

Here, the d_ϵ orbitals are closer to the ligands than are the d_γ orbitals; but neither point directly towards them. The result of this is that *the energy of the d_ϵ orbitals is raised more than the energy of*

the d_γ orbitals, but the energy difference is not so great as with octahedral complexes.

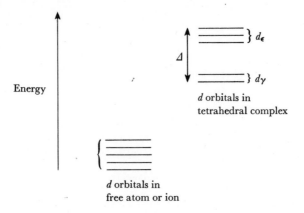

For a particular metal ion, it turns out that

$$\Delta \text{ tetrahedral} \approx \tfrac{4}{9} \Delta \text{ octahedral}$$

If we consider two Co^{2+} complexes in different environments:

$[Co(H_2O)_6]^{2+}$	and	$[CoCl_4]^{2-}$
I		II
octahedral		tetrahedral
(pink)		(blue)

There are two important differences between them:

(a) Cl^- produces a weaker field than does H_2O.

(b) The tetrahedral field is weaker than the octahedral field, i.e. the separation between the orbital subsets is smaller in complex II than in complex I. This means that light of lower energy (or longer wavelength) will be required to excite electrons in II compared with I. In complex I, light is absorbed in the blue region of the spectrum and thus produces a pink coloration; in complex II, lower energy radiation (i.e. red light) is absorbed to produce a blue coloration.

In summary, we see that the simple electrostatic theory readily explains some of the properties of transition metal compounds. There is considerable evidence available, however, which shows that the overlap of metal and ligand orbitals occurs (i.e. covalent bonding). The degree of covalent character depends greatly on the ligand, being small for ligands such as fluoride, and large for ligands such as cyanide and carbon monoxide. The differing degrees of overlap are best dealt with by using the molecular orbital theory, which is beyond the scope of this book.

CHAPTER 9

INTERMOLECULAR FORCES

We have already seen that in ionic solids the forces holding the ions together are electrostatic in nature and in order to break down these crystal lattices large energies are required. Many covalent compounds form crystal lattices and evidently attractive forces exist between the discrete molecules in the crystal. It is interesting to inquire into the nature of the forces which hold these crystals together.

In 1873 van der Waals postulated the existence of attractive forces between all atoms and molecules and that these forces are electrostatic in origin. There are two types of intermolecular force:

1. Dipolar Forces. In covalent bonds formed between atoms of different electronegativity, the most electronegative atom has the greater share of the electron pair. This results in a charge separation in the molecule and the molecule is said to possess a dipole. The dipoles of different molecules align themselves so that electrostatic attraction between opposite charges occur. This is shown below

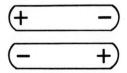

2. Dispersion forces. Even in atoms and molecules which do not possess dipoles of the above type, attraction occurs. The noble gases form crystal lattices at sufficiently low temperatures. A neutral atom such as a noble gas consists of a positively charged nucleus with negative electrons revolving around the nucleus. At any instant in time there will thus be a charge separation between the nucleus and the electrons, i.e. a dipole. The rapid movement of the electrons ensure that the charge separation is the same in all directions and therefore a physical measurement would reveal a zero dipole. However for another atom in the vicinity, dipole attraction occurs at a particular instant in time as shown below:

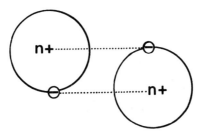

The greater the number of electrons that the atom possesses the more such interactions, and hence the greater the attractive force. There is thus an increase in the boiling points of the noble gases with increasing atomic number. This also accounts for the increases in melting and boiling points observed for series of related compounds, e.g. homologous series of paraffins and halogens.

The magnitude of van der Waals forces is much smaller than that of ionic forces, as may be seen by comparing the lattice energies of NaCl (184 kcal/mol, or 770 kJ/mol) with HI (6·2 kcal/mol, or 25·9 kJ/mol).

Chapter 10

BONDING IN METALS

In the lattice of a metal, the atoms, with few exceptions, are arranged in one of three ways as shown below:

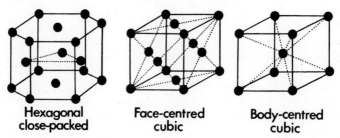

Hexagonal close-packed   **Face-centred cubic**   **Body-centred cubic**

In each case the atoms are close packed and for the first two types each atom has twelve near neighbours and the third type eight near neighbours. The forces holding the atoms together are much larger than van der Waals forces and vary over the wide range 18–200 kcal/mol (75–840 kJ/mol). Since there are many atoms surrounding an individual atom, there are insufficient valence electrons available to form simple two electron bonds between each atom in the three-dimensional lattice. An important property of a metal is that it has high thermal and electrical conductivities. This suggests that the electrons are able to move freely through the whole lattice. A simple theory which explains this and other properties is that the metal lattice consists of positive metal ions held together by electrons:

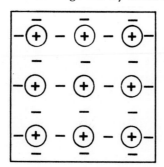

The positive ions may be thought of as being embedded in the electron cloud of all the valence electrons. This model also explains why the cohesive forces and hence the melting points of the metals vary widely. The alkali metals melt at lower temperatures than the alkaline earth metals; in the latter there are twice as many valence electrons available and therefore the ions are held more firmly together. Similarly transition metals which have many more valence electrons melt at much higher temperatures.

A valence bond description considers the bonding in terms of a resonance hybrid of the numerous electron pair bond forms, two of which are shown for lithium below:

```
Li—Li  Li        Li—Li  Li
       |                |
Li  Li  Li       Li  Li  Li
|   |            |
Li  Li  Li       Li  Li—Li
```

There are a very large number of canonical forms for the metal lattice, which leads to a large resonance energy and hence large cohesive forces. The greater the number of valence electrons available the greater the number of bonds per atom in the lattice and hence the greater the cohesive forces.

Since the metal atoms are packed close together in the lattice, the orbitals of a given atom will overlap the orbitals of several other atoms. Delocalised molecular orbitals which extend over the whole lattice may be constructed, using methods similar to those outlined previously. This method is particularly useful in explaining the properties of metals but is beyond the scope of this text.

Chapter 11

HYDROGEN BONDING

The hydrogen atom is unique, amongst the elements which form chemical compounds, in that there are no electrons beneath the valence shell, only a bare proton. Thus when hydrogen becomes bonded to another element, since the electron of the hydrogen atom spends much of its time in the covalent bond formed, another molecule in the vicinity will be exposed to the strong positive charge of the proton. The positive charge will be greater if the hydrogen is bonded to a strongly electronegative element. Then a weak electrostatic bond is formed between the hydrogen atom on one molecule and the strongly electronegative element of another molecule. This may be represented as:

$$Y\text{----}H\text{---}X$$

The weak electrostatic bond, which is shown as a dotted line, is called a hydrogen bond. X and Y are usually the most electronegative elements, i.e. F, O or N. Although these bonds are weak and have energies between 1 and 10 kcal/mol (4 and 50 kJ/mol) compared with ordinary covalent bonds (30–100 kcal/mol, or 125–420 kJ/mol) they have an important bearing upon chemical and physical properties. A few examples are given below.

It is well known that the boiling points of NH_3, H_2O, and HF are anomalously high compared to the hydrides of the other elements in their respective groups. The boiling points are shown below, together with those of a normal series, namely Group 4.

Boiling Points of Hydrides (°C)

CH_4	−161	NH_3	−33	H_2O	100	HF	20
SiH_4	−112	PH_3	−88	H_2S	−60	HCl	−85
GeH_4	−88	AsH_3	−62	H_2Se	−41	HBr	−67
SnH_4	−52	SbH_3	−17	H_2Te	−2	HI	−36

The anomalous values may be attributed to molecular association brought about by hydrogen bonding:

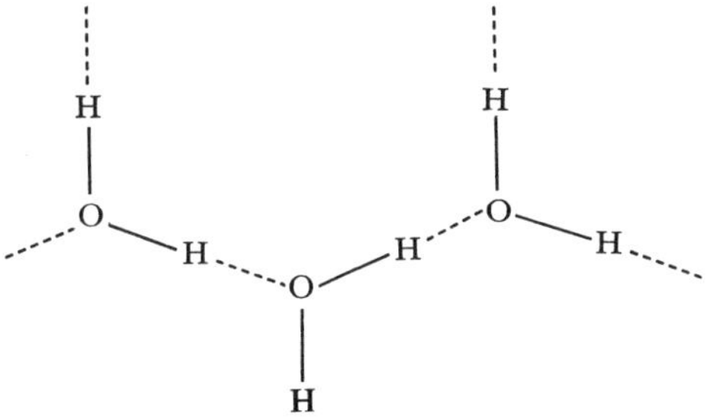

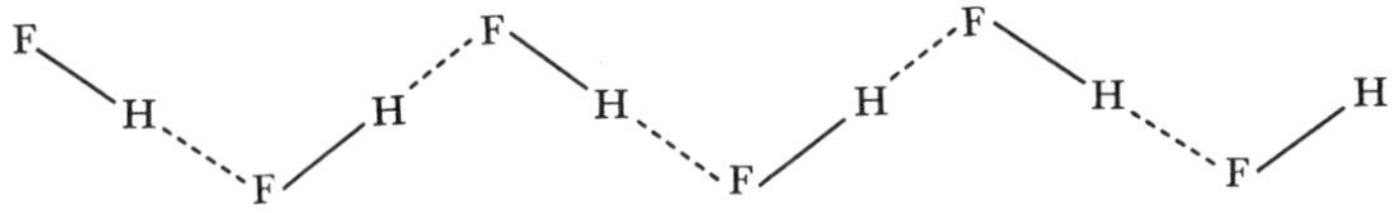

HF is associated even in the vapour phase, where chain lengths of up to six HF units are present.

Alcohols, amines and carboxylic acids are other examples where extensive hydrogen bonding occurs. Acetic acid is dimerised even in the vapour phase:

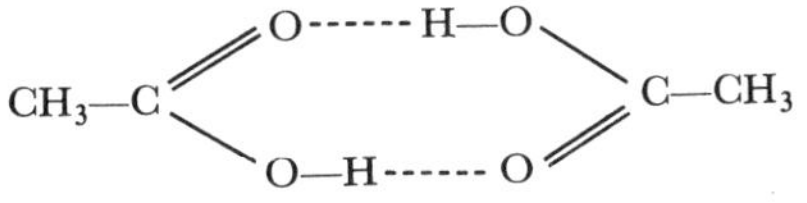

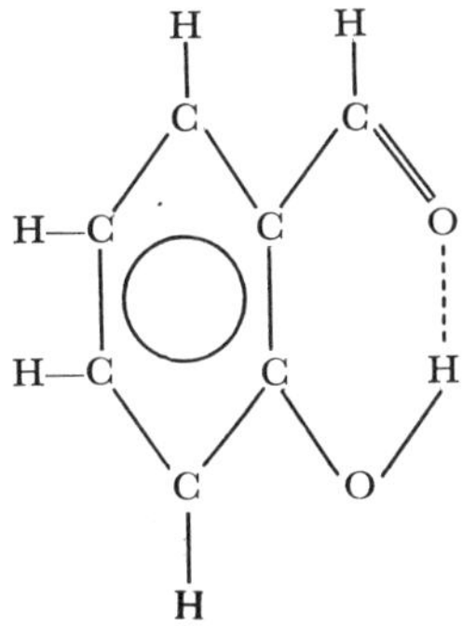

All the above examples illustrate intermolecular hydrogen bonding. In some molecules their configurations allow intramolecular as well as intermolecular hydrogen bond formation, e.g. salicylaldehyde. Intramolecular hydrogen bonding permits the formation of a six membered ring (as illustrated at the bottom of p. 99). In the meta and para isomers of this compound, ring formation is not possible and only intermolecular hydrogen bonding may occur. This results in the ortho compound being considerably more volatile than its isomers.

INDEX

acetylene, 55
actinides, 14
ammonia, 23, 33, 46, 51
angular probability, 41
antibonding, 64
antiferromagnetism, 81
atomic orbitals, 5, 7–14

benzene, 61
beryllium
  electron configuration, 11
  chloride, 25, 31, 47
body-centred cubic structure, 21, 96
Bohr theory, 2
bond, π-, 53–62
bond, σ-, 51
bond order, 72
bond strength, 72
bonding in transition metal compounds, 80–93
boron
  electronic configuration, 11
  trifluoride, 27, 32
  hydrides, 73

caesium chloride
  energetics, 19
  unit cell, 21
carbon, electronic configuration, 11
carbon dioxide, 57–59
carbonate ion, 59
chlorine fluorides, 31
colours of transition metal complexes, 90
coordinate bond, 27
coordination complexes, 82
coordination number, 83
covalent bond, 23

degeneracy, 87
delocalisation, 62
diamagnetism, 80
diatomic molecules, 63
diborane, 73
dinitrogen tetroxide, 60
dipole forces, 94
dispersion forces, 94

electron, 2
electron affinity, 18
electron deficiency, 74
electron density, 45
electron pair repulsions, 31
electronic configuration, 10
electronegativity, 37
electrostatic crystal field theory, 85
energy level, 3, 9, 63
ethylene, 54

face-centred cubic structure, 21, 96
ferromagnetism, 81
fluorine
  electronic configuration, 12
  molecule, 72

hexagonal structure, 96
Hund's rule, 11
hybridisation, 49
hydrogen
  atom, 2, 5
  bonding, 98
  fluoride, 25, 99
  sulphide, 29

inner orbital complexes, 85
intermolecular bonding, 94
intermolecular forces, 99
intramolecular bonding, 99
ionic compounds, 16
ionisation potentials, 17
  table of, 18

lattice energy, 20
ligand, 82
lithium, electronic configuration, 10

main group elements, 14
metals, bonding in, 96
methane, 26, 32
molecular orbitals, 62, 63
molecular shapes, 31

neon, electronic configuration, 12
neutron, 2
nitrogen
  electronic configuration, 11
  molecule, 56, 72

nitrite ion, 59
noble gas compounds, 76

outer orbital complexes, 85
oxygen
    electronic configuration, 11
    molecule, 70

paramagnetism, 80
periodic classification, 12
periodic table, 13
phosphorus
    electronic configuration, 28
    pentafluoride, 29, 34
    trifluoride, 28
promotion energy, 26
proton, 2

quantum number, 7

resonance, 57
    energy, 58
    hybrid, 22

sodium chloride unit cell, 22
spin free complex, 88
spin paired complex, 89
strong field, 89
sulphur hexafluoride, 30, 36
sulphur tetrafluoride, 30, 35

three-centre bond, 73
transition metals
    electronic configuration, 81
    electrostatic crystal field theory, 85
    valence bond theory, 81

uncertainty principle, 5, 16
unit cell, 21

valence bond theory, 39
van der Waals forces, 94

water molecule, 24, 34, 47
weak field, 89

xenon fluorides, 76